Preface

Automated *information systems* are a restricted class of general software systems which have a widespread presence in government, industry, and commercial enterprise. The importance of such systems in the effective running of these organisations is well accepted and the overriding goal of software engineering is to ensure that information systems meet user requirements within reasonable cost.

As a consequence of their importance, the development of information systems has attracted considerable attention from researchers, practitioners and software and hardware vendors; the results of which have been the improvement of database technology, programming languages, operating systems and hardware.

Underlying these developments has been a growing awareness that two factors are essential in the effective development of information systems: the need for a systematic process of development and the importance of adequate requirements specification and design. The first of these factors is concerned with the management of the development process and co-ordination of large development teams. The second of these factors is concerned with the correct capture of user requirements and their representation in a system whose structure matches the structure of the users' problem definition. These two activities give rise to the discipline of *Systems Analysis and Development* which is the subject matter of this book.

The book begins with a general examination of the issues of organisational systems and the role of information within such systems. The requirements of the development process are also introduced in chapter 1.

Chapter 2 begins with an introduction to the traditional approach to systems development and discusses the challenges to this approach which have led to the emergence of contemporary system development methods.

Appendix A presents a comprehensive list of these methods and a cursory glance suggests that with the large number of methods currently used, there is no single accepted approach to information systems development. We therefore argue that there is a need to develop a framework within which the reader can explore the different development philosophies and formalisms currently in practice. Chapters 3 to 9 detail the models and techniques that lie within the general framework outlined in chapter 2.

Chapter 3 begins with a review of the issues and basic stages of problem analysis, in which analysts seek to define the boundary of users' problem domain and identify the key information requirements and processes. The main fact gathering and documentation techniques are also reviewed.

Chapters 4 and 5 present the complimentary activities of process modelling and conceptual data modelling. In chapter 4, the techniques of user task modelling, data flow analysis and process description are described, whilst chapter 5 addresses the issue of conceptual data modelling. In this edition of the book more emphasis has been placed on the concepts and techniques associated with the development of a conceptual schema which is independent of any implementation considerations and can be mapped onto database structures. The material presented in chapter 5 begins by introducing the conceptual schema and establishes a set of desirable characteristics of conceptual modelling formalisms. Subsequently, a set of basic concepts underlying two conceptual modelling formalisms are defined and highlighted through the use of numerous examples.

Chapter 6 represents a turning point in the material of the book, moving away from the issues of systems analysis and onto the issues of system design. The material begins with a review of validation techniques, such as walkthroughs, reviews and inspections. This is followed by a series of sections outlining specific design issues such as overall system architecture, security, safety and operational structure.

Chapters 7 and 8, correspond to chapters 4 and 5, but at the design level. Chapter 7 takes a dual approach to process design by examining process-driven and data-driven approaches. In the former case, the concepts of coupling and cohesion are presented, together with transform and transaction analysis. With the data-driven approach, the principles of data structuring and correspondences are explained. Chapter 8 continues an examination of design issues, but this time from a data perspective. Issues of file design and database schema design are examined and the design of record structures is examined in the context of two techniques: data normalisation and direct mapping from a conceptual schema.

In this edition, the material in chapter 9 has been extended to incorporate some of the principles of dialogue modelling that are found in the emerging methods and support tools. The chapter also retains previous coverage of screen and report design, together with an explanation of interfacing issues relating to hardware selection, code design and data validation techniques.

Chapter 10 examines some of the less technical issues involved in system development, including the introduction of a new system and overall project management.

The material in the book is supported by many references, often seminal papers on a particular topic and it is hoped that those readers who wish to pursue further reading of individual areas will find these references of use. The book also contains two appendices. Appendix A gives a comprehensive bibliography of the most widely practiced information systems development methods. Appendix B introduces a case study which has served as a common basis for demonstrating many of the concepts and techniques covered by this book.

Finally, the authors would like to express their thanks to their colleagues, students and readers of previous editions of this book for the valuable comments they have made and in helping to produce this edition.

Manchester,
February 1989.

Contents

Chapter 1

Information Systems

The force behind the effective functioning of any organisation is a system. Organisations consist of many parts which can themselves be regarded as systems and whose interaction is a prerequisite for the efficient functioning of the organisation. This book is concerned with one particular type of system found in organisations, *information systems*, and the way in which such systems are developed.

This chapter begins by considering the wider issues of systems as applied to organisations. A model for understanding organisations is described and the role of information systems within this model is discussed. Finally, the steps for developing computer-based information systems are briefly described.

1.1 Organisational Systems

The term *system* has many interpretations, but in its simplest form it can be regarded as a collection of interrelated parts which act as a whole towards a common goal. This definition is a convenient way of classifying concepts and phenomena in widely diverse fields and is particularly useful when examining organisations.

Externally, organisational systems have boundaries, beyond which exists an environment in which the system operates and because of this interaction organisational systems are regarded as *open* systems. Internally, systems will contain a number of subsystems which, through interaction, serve the common purpose of the system. This interaction is governed by a set of rules and stimuli from the external environment. For example, consider a company which manufactures, imports and distributes optical frames. Figure 1.1 shows the relationships and information flows between some of the subsystems, together with the interaction between the system and its environment.

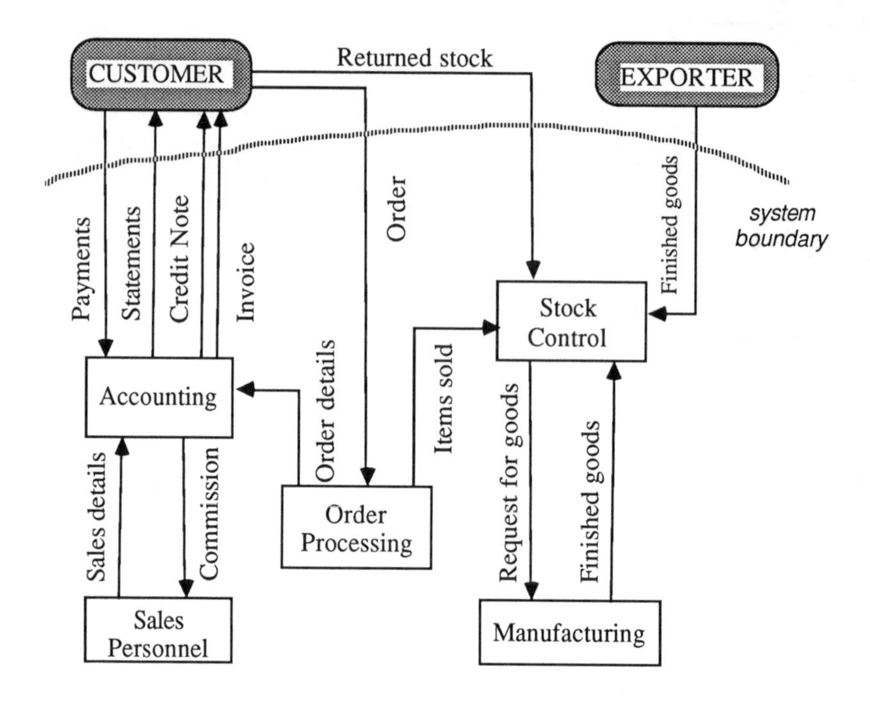

Figure 1.1: An Information Flow Diagram

The definition of a subsystem is the same as that for a system the only difference being in the particular viewpoint adopted by an observer. For example, the subsystem *accounting* in figure 1.1 may be regarded as a system in its own right if we are only interested in accounting and its activities. The terms system and subsystem are used interchangeably in this book and in particular reference is made to *information systems* within organisations when strictly speaking they should be referred to as information subsystems.

The parts of an organisational system (its subsystems) are open, i.e. they interact between themselves, if they are to be of any real use. These parts may be tightly coupled, in which case information or material is passed across the interfaces immediately it is generated. However, in many cases it is more convenient to delay this interaction and thus reduce the direct dependence between subsystems. This reduction in direct dependence is known as decoupling but it should be noted that excessive decoupling may result in inefficient systems. On the other hand tightly coupled systems are sensitive to fluctuations in performance in any of its subsystems. Therefore, a mechanism is needed which will control the interaction between an organisation's subsystems in an optimum way.

In figure 1.1, customer and exporter are regarded as being outside the business system, but they relate to the company through simple inputs such as a customer order. The arrival and processing of this input will activate a number of information flows between some of the business' subsystems. For example, details of the order will be passed to the accounts subsystem where an invoice will be prepared and sent to the customer. Similar details will also be passed to the stock control subsystem for amending the stock and so on.

In order to understand the need for information systems, we need to examine organisational structures from the viewpoint of planning and control, as well as that of decision-making.

An organisation may be viewed through the classification of its management structure and its planning and control can be divided into three levels: strategic planning, management control and operational control (*Anthony, 1965*).

Strategic Planning. Strategic planning is concerned with the setting of objectives in the organisation and the resources required for attaining these objectives. This management process is concerned with long term activities.

Management Control. Management control is directed towards medium term activities and ensures that the resources are obtained and used in an optimum way.

Operational Control. Operational control is concerned with short term activities and assures that specific tasks are carried out efficiently. It is transaction oriented, the problems are repetitious and are well structured.

To demonstrate these three levels consider the optical frames company again. The decision to introduce a new line of product would fall into the domain of strategic planning. Activities about monthly production schedules, re-organisation of manufacturing procedures etc. will be the concern of management control and, the way daily production is handled would be operational control.

The organisational model defined so far includes only the management structure. Obviously there are activities, decisions, resources etc. in organisations that are not considered as part of the management structure but are part of the operational level. At this level, inputs arriving from the system's environment are physically transformed by the organisation's resources into outputs; for example, raw material is used by machine tool operators to manufacture optical frames which are to be sold to customers.

The three level model based on the management structure can therefore be extended to give a more complete picture of organisational systems by including the operations level (*Davis, 1974*).

The discussion so far has concentrated on examining organisations from the point of view of planning and control. Another view orthogonal to this is a model based on the decision-making process. *Simon* (*1960*) in his individual work and in his joint research with Newell (*Newell & Simon, 1972*) established the foundations of human decision-making models as a three stage process.

The Intelligence Stage. This stage is concerned with the identification of a problem and the collection, classification, processing and presentation of data.

The Design Stage. This stage is concerned with planning for alternative solutions. If available data is insufficient for evaluating the different solutions then more data may be required from the intelligence stage.

The Choice Stage. A choice is made from one of the alternative solutions. The decision maker is faced with a number of difficulties during this stage. Examples of these difficulties are conflicting interests, uncertainty as to the outcome, multi-preference when having several variables not all of which are comparable etc.

Organisations have been viewed so far in terms of two orthogonal views: their management structure and the decision making process which is carried out within this structure. The activities involved in these two views are complex and difficult to carry out, without the help of one or more systems specifically assigned to supporting these activities. These are known as the *information systems*. Traditionally, one such system was sufficient for dealing with the information requirements of an organisation. However, many contemporary organisations display characteristics which may necessitate the presence of more than one information system, usually interacting with each other.

The importance of information systems in overcoming the complexities of contemporary organisational requirements and the ever increasing demand for accurate and timely information is undisputed- information being regarded as important as labour and capital in most large organisations.

The consequence of this is that the management activities within an organisation must address not only traditional management problems, but also the issues relating to the development and maintenance of its information base, in an integrated fashion.

The key question is how should an organisation manage its information resource and supporting systems? For whereas management techniques applicable to labour and capital have been practiced for many years, the importance of properly managing the development and use of information systems is only just becoming apparent.

Empirical studies show that there is no consistent pattern in achieving these objectives and responses to these problems span from piece meal development to coherent long term strategies (*Hirscheim, 1981*). Hirscheim identifies three basic management styles to managing information systems.

Technical Static Approach. This approach involves uncoordinated activities with little attention being paid either to the applications which must be developed or the way that development must proceed.

Tactical Bottom-Up Approach. In the bottom-up approach, plans for coordinating and integrating an information system are established either by individual interest groups or from within a service function. Land (*1980*) observes that such an approach leads to inconsistencies between corporate planning and information systems planning.

Strategic Top-Down Approach. This management style incorporates an information management plan into the overall corporate plan.

What is obvious from current practices is that no ideal management model to information systems has emerged as the definitive model and it is argued that further research is required in establishing the most effective style of management (*Land, 1981*).

1.2 Information Systems

1.2.1 Information Systems in Organisations

It is clear from the diagram in figure 1.1 that a system component cannot exist in isolation but needs to interact with one or more components and with the system's environment through the flow of information. Information is any form of communication, formal or informal, and is a key concept for the cohesive behaviour of a system. The generation of information and its flow through an organisation's subsystems is a prerequisite for the efficient support of the operational structures, management and the decision-making processes defined in the previous section.

At this point, it is useful to make a distinction between operational information and management information.

Operational information, which is used routinely and allows the organisation to carry out efficiently its daily functions. For example, a calculated payroll, customer invoices, telephone bills and schedules for distribution may all be operational information.

Management information, which supports the process of decision-making at the management levels and is less routine than operational information.

Obviously, the generation and flow of information in an organisation needs to be undertaken in an orderly fashion. The mechanism which enables an organisation to achieve this, is known as an information system.

An information system is defined as the mechanism which provides the means of storing, generating and distributing information for the purpose of supporting the operations and management functions of an organisation. Such systems may be manual or, more commonly they represent an integration of manual and computer-assisted components.

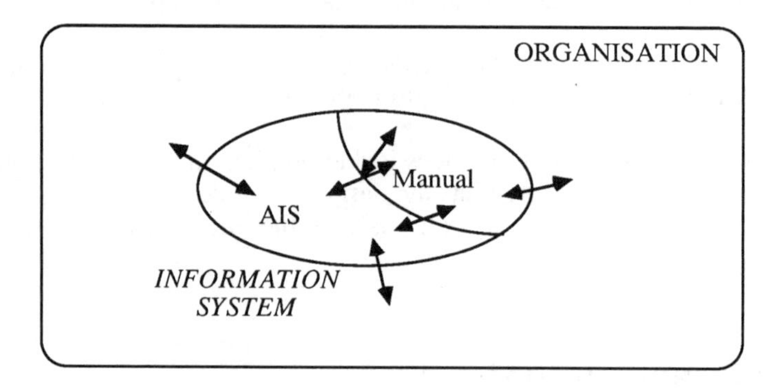

Figure 1.2: Information Systems in Organisations

This view is summarised in figure 1.2. An organisational system consists of a number of interrelated subsystems whose functionality is determined by the goals and objectives of the organisation. One or more of these subsystems is responsible for servicing the information requirements of the organisation, the *information system*. Part of such a subsystem will be *manual* whereas another will be based on some computer resource, the *automated information system* (shown as AIS in the diagram). Notice that these two parts interact with each other as well as with the organisational environment. This interaction refers to the stimuli and responses received and generated by the information system.

1.2.2 Information Systems and the Universe of Discourse

Over the past two decades two approaches have dominated work in the field of information systems: organisational mapping and organisational control (*Verrijn-Stuart, 1987*).

Organisational mapping is based on the assumption that all elements of importance in an organisation can be mapped onto a recording system, from which information can be extracted. Organisational control assumes that an organisation avails itself of information which is based on recent and past observations of real world phenomena and decisions are carried out on the basis of this information. Both mapping and control make reference to an information system.

The area of information systems is dominated by references to the real world and it has been argued that the problems found in this area are a mixture of *empirical*, *formal* and *engineering* problems (*Verrijn-Stuart, 1988*). Empirical problems are concerned with the fact that in developing information systems one is constantly engaged in observing real world phenomena. The behavioural characteristics of the optical company for example is a subject that one may investigate empirically and develop theories about it. In this sense the investigator gains *knowledge* about the optical company. Formal problems are concerned with the abstraction, structure and representation of this knowledge in a way which is possible to reason about this knowledge. Engineering problems arise when one attempts to implement the construction established by the adopted formality principles.

A corresponding view is found within the knowledge engineering field. Brachman & Levesque (*1986*) view a Knowledge Base Management System in terms of three levels: the knowledge level; the symbol level; and the system engineering level.

A definition of information systems, which recognises this three-levels view, is one which makes reference to the *universe of discourse*. This term is used to refer to that part of the real world which is of interest to a specific information system and this may be a concrete object, such as inventory system or an abstract phenomenon such as an economic system. In this way an information system is a formal description of an abstract model of a piece of reality. This description may change with the passage of time according to changes in the universe of discourse itself.

The interaction between a real system and an information system is in the form of sets of elementary facts. Obviously, these elementary facts need to be stored somewhere within the information system so that information can be used. This gives rise to the well accepted data-centered view of the

architecture of information systems (*Bubenko, 1980*) in terms of three components, the database, the conceptual schema and the conceptual information processor.

A *database* is a collection of all relevant information which describes the entities that are considered to be of interest in the universe of discourse at a specified instant or period of time.

A *conceptual schema* is a set of rules describing which information may enter and reside in the database. The information which passes from real system to information system and vice versa can be analysed in terms of the conceptualisation process (*Schank, 1975; Schank, 1976*) which regards information as being transformed into sentences, sentences transformed into elementary sentences and elementary sentences into object-role pairs. Therefore, a conceptual schema holds descriptions of object types, the roles which an object type may play in an elementary sentence and restrictions that apply to a certain object type in a certain role.

A third component of an information system is its *conceptual information processor*. The function of this component is to keep the populations of the database in accordance with the conceptual schema. Each time a request to change the database is made this component checks to see whether any of the definitions in the conceptual schema are violated and for this it needs to consider both the conceptual schema as well as the current database.

It should be noted that despite the clear distinction between the three components, in practice these components are implemented in an overlapping fashion. For example, constraints imposed by a conceptual schema may be defined partly in data structures, partly in data type definitions and partly in application programs.

1.2.3 Data-Intensive Transaction-Oriented Information Systems

In general, the term *information system* is attached to any manual or automatic system (or a combination of the two) whose function is to manage information for the purpose of decision making. However, the discussion in this section restricts information systems to a special class namely, data-intensive, transaction-oriented systems. This book is concerned with the development of automated data-intensive, transaction-oriented systems and their communication with the real environment. Such systems have a widespread presence in a variety of problem domains and can be found in most industrial and commercial enterprises as well as public organisations.

Data-Intensive systems are those systems which deal with operational, *persistent* data which is *integrated* and *shared*. Persistency implies that the data is operational data which deals with the activities of the enterprise and therefore needs to be kept for some period of time. Integration means that the total collection of data may be considered as a unification of several otherwise distinct smaller collections. Sharing means that individual pieces of the integrated collection of data may be shared by many different users.

Transaction-Oriented systems are those systems which deal with transformations on the state of data as a response to some happening in the organisation. For example, the reservation of a room in a hotel would be reflected in an information system as a transaction which changes the state of the data pertaining to the specific room from being *available* to *booked*.

Underlying the operation of data-intensive transaction-oriented systems- henceforth simply referred to as information systems- is the notion that each data transformation must exhibit the properties of consistency, atomicity and durability. Consistency means that a transaction must obey certain legal protocols; atomicity means that the transaction either happens or it does not; and durability means that once a transaction has been committed it cannot be abrogated (*Gray, 1981*).

Therefore, an information system may be viewed as an abstract model of three layers as shown in figure 1.3.

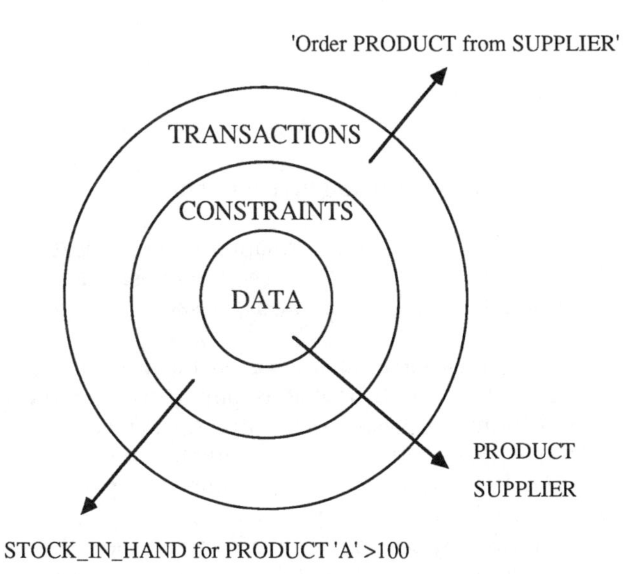

Figure 1.3: Data-Intensive Transaction-Oriented Systems

The position of each layered component in the diagram of figure 1.3 is significant; the diagram is meant to imply that there is an inwards dependency of layers. At the centre of any information system lies the data which describes the universe of discourse at any point in time. The state of the collections of data is governed by a set of constraints whose function is to ensure the integrity of all data between transitions from each state to its successive state. Finally, the operations which determine each transition are governed by transactions.

1.3 Developing an Information System

An information system, as shown in figure 1.2, resembles a *black box* and says very little about the workings of such a system. However, two aspects of information systems are of importance to students of Computer Science: "what is the technology which makes an information system work?" and "how are information systems developed?". The first question is concerned with issues such as databases, data communications, programming languages etc, issues which are dealt in detail by companion books in this series. The second question relates to the process of developing software which constitutes the automated information system of an organisation and it is a question which is directly addressed by this book. To this end, development of an information system is regarded in this book as a process of developing a system's:

- transactions
- data and its constraints
- interaction with the surrounding environment.

The requirement for developing an information system arises from the need for change through an interaction of three variables: the technology, the operational needs and the management needs, as shown in figure 1.4.

Technological innovations have an impact on the way that technology is perceived by organisations and how it is put to use for the purpose of developing new information systems. Historical evidence shows that it is possible to identify distinct invention, innovation and diffusion cycles (*Marchetti, 1981*). It is argued that inventions in a particular discipline follow a wave pattern, the invention cycle is followed by an innovation cycle which in turn is followed by a technology diffusion cycle. Observations over the past 300 years show that the time taken for a cycle to be completed is diminishing. In computing terms these cycles are accelerated at an ever increasing rate. Technological advances in hardware, telecommunications, databases, expert systems, office systems and in many other areas has meant that organisations consider the introduction of some new technological innovation approximately every five years.

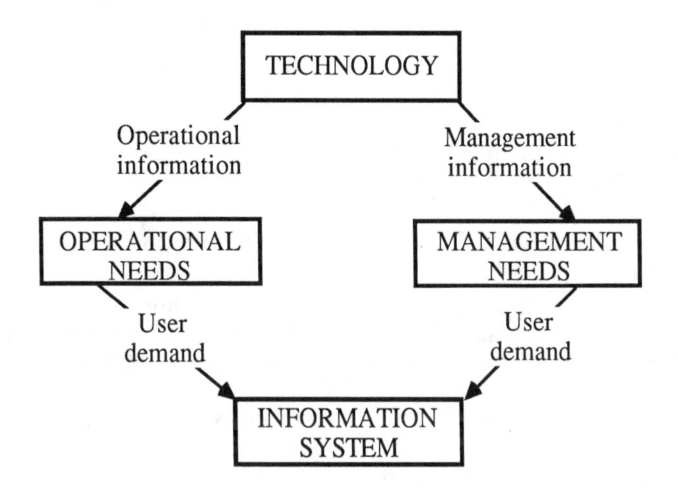

Figure 1.4: Influences to Changes in Information Systems

The ever increasing complexity of organisational systems has meant that, in order for the organisation to function properly, there is an increased demand for improved (or very often new) information systems. Similarly, effective management requires relevant, timely and accurate information and this is another contributory factor to the demands for new or improved information systems.

The complexity of contemporary information systems necessitates the need for a methodical approach to the development of such systems and to this end a variety of approaches exist. Various software engineering paradigms have been proposed, from the *lifecycle model* (*Royce, 1970*) to *prototyping* (*Boehm et al, 1984; Mayhew & Dearnley, 1987*), *fourth generation techniques* (*Cobb, 1985*) and *formal* approaches (*Jones, 1981*).

The classical model of information systems development is the *lifecycle model* whereby a software system emerges by following a well defined set of interrelated activities carried out by development personnel, together with the ultimate users of the system. (It is worth noting that although many system development lifecycles are presented as a sequence of activities, most processes of software development are considerably more complex and may involve iteration during development phases).

The objective of the lifecycle approach is to deliver robust easily maintainable systems which will last as long as possible. Within the systems development lifecycle the following are generally regarded as forming a set of interrelated activities:

- *requirements definition*, in which the purpose, functional requirements and constraints of the proposed system are identified
- *design*, in which plans are developed to show how the proposed system will be implemented
- *implementation and testing*, in which the proposed system is coded and implemented
- *operation and maintenance*, in which the new system operates and is modified where errors are discovered or requirements change.

The lifecycle approach is a management tool which is used to plan and control the information system development effort. The techniques involved are concerned with the identification and specification of deliverables which constitute documentation of one sort or another. This documentation is usually formally reviewed, by the developer and the acceptor, at different milestones which are themselves determined by the tasks identified within the key development activities outlined above. The term *developer* is used here as a composite term to refer to computer specialists including systems analysts, systems designers and programmers. On the other hand *acceptor* refers to the end-users, managers, clerks, etc. who are affected by the introduction of a new system. Very often, systems analysts together with end-users form a steering committee whose function is to monitor the progress of the system under development.

A different approach, but often complementary to the lifecycle model, is that of *prototyping*. The objective of prototyping is to produce a *rough and ready* working model of the planned component of an information system as early as possible, verify this with the end-users and implement any changes which may be necessary from such a consultation. To achieve this objective a developer will make use of software support facilities such as application generators, report generators, query languages and graphics interfaces.

The impetus behind prototyping is the need for bridging the communication gap between developers and end-users at an early stage of development. This gap exists for a number of reasons, but it is usually because end-users perceive a problem differently to the way developers view the same problem or because end-users are uncertain about the planned information system. Prototyping attempts to remedy both of these problems.

The use of fourth generation techniques has proved to be useful for problem areas which have a well identified boundary and are primarily interactive in nature. These techniques encourage a developer to first capture the users' requirements and go directly to implementation using a fourth generation language. Finally, formal methods are concerned with the correctness of software, starting from a point where a requirements specification has been established.

Without excluding prototyping or any of the other approaches as being of importance to information system development this book adopts a structured way of examining the concepts, techniques and tools associated with such an undertaking by adopting a framework closely resembling that of the lifecycle model. Within this framework two distinct stages are recognised:

The Expansion Stage, in which a system is decomposed into its subsystems and their interrelationships analysed.

The Contraction Stage, in which the constituent parts of a system are synthesized into a new system.

Within this framework it is possible to view an information system as a model of some slice of reality about an organisation, the facts which exist in the organisation and the activities which take place. Therefore, the problem of developing an information system may be regarded as a problem of model description. These models fall naturally into two categories dealing with the dynamic and static aspects of an information system.

Dynamic aspects are concerned with events, flows, actions, transactions and control structures. They can be regarded as the usage perspective i.e. they reflect 'how the organisation operates'. Static aspects are concerned with objects, categories and associations, the information structure perspective i.e. they reflect 'what supports the organisational operations'. Dynamic aspects and static aspects, together with the way they interact with each other, must be modelled first and then translated into program and data structures, respectively.

The activities which occur during this lifecycle, span over two realms, the conceptual realm and the implementation realm and these are summarised in figure 1.5.

For the *conceptual* realm the systems analyst will use models to abstract and understand the facts and rules of a system, which correspond to its static and dynamic aspects, respectively. The resulting models should be independent of any physical considerations.

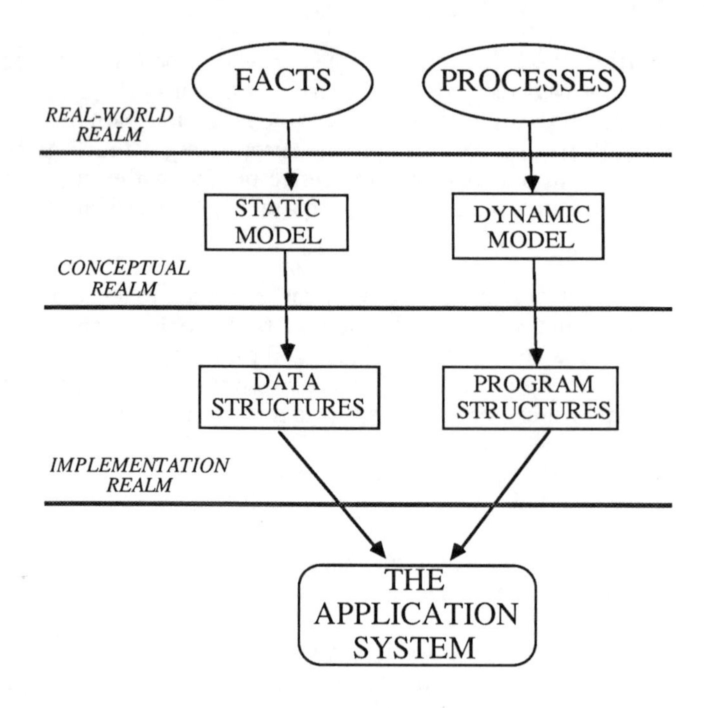

Figure 1.5: Static and Dynamic Aspects of an Information System

For the *implementation* realm the designer will use the conceptual models and will attempt to provide the facilities required by the users taking into consideration certain constraints imposed by the target physical architecture.

A simple view of the process of developing software for an automated information system is shown in figure 1.6.

This view recognises the importance of first developing models which are oriented towards the understanding of the application domain (i.e. the models are cognitive in nature) and then through a series of transformations, developing more formal models which are guided by physical considerations.

In practice, there are a number of phases involved in developing a computer-based information system and these are described in the following sections.

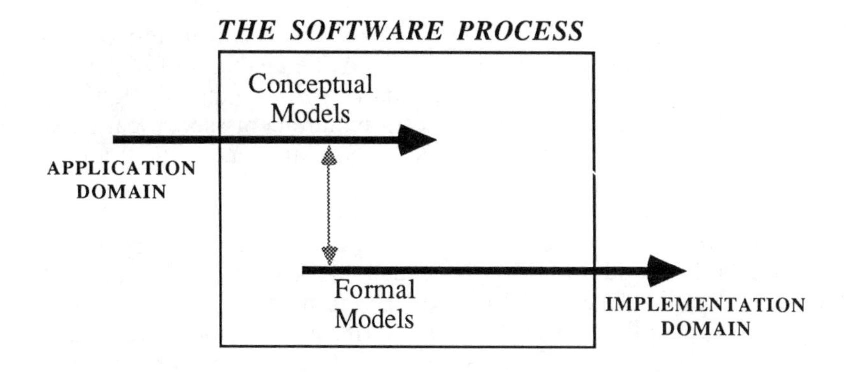

Figure 1.6: A Simple View of Software Development

1.3.1 The Information Strategy Planning Phase

The information strategy planning phase can be regarded as having two, interconnected aims. Firstly, it seeks to develop an overview of all parts of an organisation, in terms of the organisation's objectives and related information needs. Secondly, this phase aims to establish the scope of the system investigation and to draw up terms of reference for further work. It is assumed that initial meetings with management have taken place, a preliminary feasibility study has been conducted and the *go ahead* decision has been made. The detailed objectives are:

- to identify system goals and the scope of the system development
- to define a technical architecture which provides a statement of direction for hardware and software facilities
- to outline proposed arrangements for management and control of the information systems activity within the organisation
- to formalise a problem definition which will act as a control mechanism for the remaining phases of the development process.

1.3.2 The Requirements Analysis Phase

During this phase the systems analyst is concerned with the identification and modelling of the important elements of the system. The approaches adopted during this phase have been termed *requirements engineering*.

Requirements engineering (*TSE, 1977; COM, 1985*) is concerned with the application domain and its objective is to provide the basis of understanding real world phenomena and the users' (non-computer experts) requirements. This is achieved by the development of a requirements specification which states the desired characteristics of a system component without any actual realisation in computing terms.

According to *Rzepka and Ohno* (*1985*) a requirements specification represents both a model of what is needed and a statement of the problem under consideration. The model is developed through a process of analysing the problem, modelling the results and verifying the model with the users for accuracy. *Dubois et al* (*1986*) argue that within requirements capture two basic activities take place: modelling and analysis. *Modelling* refers to the mapping of real world phenomena into basic concepts of the requirements modelling language. *Analysis* refers to techniques used to facilitate communication between requirements engineers and users using the conceptual schema as the basis of this communication.

It has been argued that the effectiveness of information systems is greatly enhanced if a developer adopts some *conceptual modelling* formalism. Conceptual modelling encourages the developer to concentrate on the semantics of the application domain rather than the characteristics of the delivered system. It is argued that models oriented towards the description of the problem domain should be independent of any implementation considerations and should concentrate on the behaviour of the modelled component (*Balzer & Goldman, 1979*). Conceptual modelling is a natural and effective way of organising real world facts and the models created by this process can be mapped onto machine level models at some later stage quite effectively. Furthermore, because of their closeness to the human perception, conceptual models greatly enhance the communication efforts of the systems analyst with end-users.

Conceptual models represent abstractions, assumptions and constraints about the application environment (*Bubenko, 1980*). More accurately these models represent the user's conceptualisation of the application environment.

In summary, the objectives of this phase are:

- to accurately model the part of the system in which the analyst is interested
- to accurately model the user requirements
- to encourage user involvement
- to produce an analysis specification which can be transformed easily

into a design specification

- to fully document the existing system

- to co-ordinate the work of many analysts working on a large complex system and consider many user views, thus resolving conflicts, indeterminancies, redundancies etc.

- to view the system as a *whole* i.e. a set of interrelated parts which all act together towards a common goal.

1.3.3 The Design Phase

The design phase is concerned with the construction of the proposed system which is done at two levels. The logical level which is independent of any target hardware and software environments; the physical level which refers to the design model on a particular hardware configuration with its available resources and imposed constraints.

The physical model is a compromise solution between many conflicting factors, for example, cost, expandability, reliability, performance etc. This means that there is no single path from the logical design to the physical design and an analyst will present end-users with a variety of options. The physical design will depend on the decision of the system's acceptor. The objectives of system design are:

- to transform the requirements specification into logical and physical models

- to provide a specification for man-machine interface

- to evaluate design

- to fully document design model and enforce maintenance of documentation.

1.3.4 The Implementation Phase

During the implementation phase the physical model, constructed during the design phase, is transformed into a working computer-based information system. This may be achieved through a variety of approaches depending on the type of the target system and the available resources. For example, consideration must be given to the programming environment (traditional programming versus application generators), the data management system (filing versus databases), the user interface, the processing mode (batch, on-line, distributed) and so on. In summary, implementation is concerned with three key issues:

- how to create a physical database as specified in the design model
- how to transform the designed system into executable code
- how to carry out the changeover procedures in order to start operating the new computer system.

1.3.5 The Maintenance Phase

The maintenance phase is concerned with the development of a system after it has entered the production stage. Three factors may cause the need for this further development and these result in three standard types of maintenance (*Lientz and Swanson, 1980*).

Corrective Maintenance. This type of maintenance activity arises where the system currently in operation does not meet the stated user requirements. This can result from a number of causes and include incorrect capture of initial requirements, poor design or bad implementation. Maintenance is generally regarded to fall into this latter category, although it is increasingly recognised that incorrect requirements capture is a greater problem.

Perfective Maintenance. Typically, as experience is gained with a new information system, certain inefficiencies within its operation will arise. Perfective maintenance is concerned with the improvement of a system, without affecting its basic functionality. Examples of such maintenance would include improving file accessing, using new algorithms to speed numerical computations and similar improvements.

Adaptive Maintenance. Whilst corrective and perfective maintenance is extremely important, they are essential short-term issues. Over the longer term, the initial user requirements for a system will change. Organisations exist in a real world which is changing. This in turn will cause the business requirements of an organisation to change and consequently changes to its information needs. Adaptive maintenance is therefore concerned with the *evolution* of an information system and changing its underlying functionality to meet additional or changed requirements.

1.4 Summary

Contemporary organisational systems are large and complex. The need for integrated decision making has highlighted the need for integrated information systems. Such systems need to be supported by a computer and their development requires a considerable human effort so that the delivered system can be efficient and flexible.

An information system can be viewed as a model of some slice of reality about an organisation, the facts which exist in the organisation and the activities which take place (*Borgida, 1986*). Therefore, the problem of developing an information system may be regarded as a problem of model description. These models fall mainly into two realms: the conceptual realm and the implementation realm.

Information systems development has progressed from the ad-hoc piecemeal approach which paid attention only to the models in the implementation realm, to more disciplined approaches which have shifted the emphasis towards the conceptual realm. At the same time these approaches have emphasised the need for a development discipline. The main reason for a development discipline arises from the fact that the requirements and consideration of present-day, large, complex systems are beyond the full understanding by one person (*Lehman, 1978*). This necessitates a team approach to the development task and thus the need for well defined steps, in order to ensure proper co-ordination and control of teams. Disciplining the software development process is an essential prerequisite for managing changes to software (*Bersoff et al, 1979*).

Chapter 2

Approaches to Information Systems Development

This chapter is concerned with the approaches available to analysts and designers for developing information systems. The emphasis of this chapter is on the concepts and techniques propagated by contemporary development methods.

As an introduction to the themes explored in this chapter, traditional techniques are considered first. However, the traditional approach has been criticised as being inappropriate for today's complex information systems and a number of different development methods have been proposed.

Currently there exists a large number of such methods and this chapter attempts to classify them according to their underlying philosophy, approach and modelling orientation. Such a classification is both useful and necessary if some generalised and unifying principles can be derived in order to objectively discuss the concepts and techniques adopted in developing information systems. A framework is introduced in this chapter which serves as the basis of describing the process of information system development in subsequent chapters.

2.1 Traditional Approaches to Development

The traditional approach to system development was piecemeal, in which applications were designed independently of each other. By contrast, today's systems are complex and their high degree of interdependence requires an integrated solution. Furthermore, today's computer applications need to cover not only operational-level systems but also systems for tactical and strategic management.

Thus, the increased scope and sophistication of today's systems necessitate methods for developing such systems in radically different ways to those used for ad-hoc development. Unfortunately, improvement in system development techniques has not kept pace with the improvement in computing hardware. There is a long lead time before any new approaches are adopted by system developers (*Couger et al, 1982*). Consequently many computer-related applications are still developed using one or more of the traditional techniques but, as will be discussed, these techniques are rapidly being replaced by a new generation of development approaches. However, for historical reasons and because some traditional techniques are still practiced, this section gives an overview of these.

The underlying philosophy of traditional techniques is their concern with the identification and documentation of the flow of work through an organisation and the sequence of activities for achieving system objectives. No guide is given as to how these techniques may be used, instead an analyst uses his experience and know-how to derive a system specification. This specification is documented in the form of charts which in turn are supplemented by narrative.

Traditional techniques owe much to the way manual systems were observed and documented before the arrival of the computer. With the arrival of the computer and its application to business systems, analysts continued to use the same techniques usually with some minor modifications to suit the new type of usage. The most popular of these techniques are those of:

- the forms flowchart
- the clerical procedure flowchart
- the system flowchart
- the flow diagram.

The common denominator for the majority of these techniques is the concept of the *flowchart*. Process flowcharts were used by industrial engineers in the early 1900's. However, these flowcharts were used for very different purposes to those applied later on for software development; they showed the flow of physical products through a production process. Analysts modified these techniques to suit their purpose of carrying out computer-related systems analysis. The flowcharting techniques used in documenting computer systems evolved from techniques such as the process flowchart and the multicolumn flowchart.

The *process flowchart*, one of the earliest techniques, is a diagrammatic notation using special symbols to depict operation, transportation, inspection, delay and storage. The use of the process flowchart reflects its use in industrial engineering environments. Materials, operations, inspections and timing are the main considerations in drawing a process flowchart.

One of the main criticisms levelled at the process flowchart is that it does not show activities performed by individuals. The *multicolumn process flowchart* was developed to show what each worker did at each stage in the production process. The impact of the multicolumn process flowchart in analysing systems is in its ability to show multiple activities and cycles of process flow.

The *forms flowchart* is a direct descendant of the process and multicolumn charts and is used to depict the flow of forms in a system. A small number of special symbols are used to identify different types of form supplemented by vertical lines to identify different organisational responsibilities for various activities.

The *clerical procedure flowchart* shows the flow of control in the way an activity is carried out in the organisation. There is a small set of symbols used which are concerned with the initialisation and termination of an activity, the definition of procedures, and the definition of decisions. Normally, a clerical procedure flowchart is drawn as a multicolumn chart where each column represents one department.

The *system flowchart* provides an overview of the processing which takes place at the computer end. It shows what is done by the system at a high level in terms of the input data, the data processing activities and the output from each step. Although a system flowchart is concerned more with the implementational characteristics of a system the technique is often used during analysis. The consequence of this is that a system development is driven by what the physical computer system is likely to be rather than on a real understanding of the problem and the user requirements. A system flowchart is often supplemented by a *flow diagram* (also known as a program flowchart) which describes the defined system in greater detail. Because of the great detail shown in a flow diagram it is often necessary to identify connections between pages and activities. It is not unusual therefore for a flow diagram to run over many pages.

Figure 2.1 shows an example of the system flowchart using also the multicolumn approach. In this case the columns are standard corresponding to the input (I/P), master files (M/F), processing, transaction files (T/F) and output (O/P).

In this example, the system flowchart shows the major components of a payroll system. The input is the clock card data which is validated with an exception report generated for all invalid data. The valid data is stored, by the validation program in a transaction file. This file is in turn sorted in the same order as the master payroll file, producing a sorted valid transactions file. Following this, the major processing takes place i.e. the payroll updating program is executed. The updating program requires the previous payroll personnel master file (B/F), and by traversing the transaction file all corresponding records in the master file are updated and the new master file is carried forward (C/F). The update program generates the monthly payroll transaction file which is input to the print program for the production of the pay cheques.

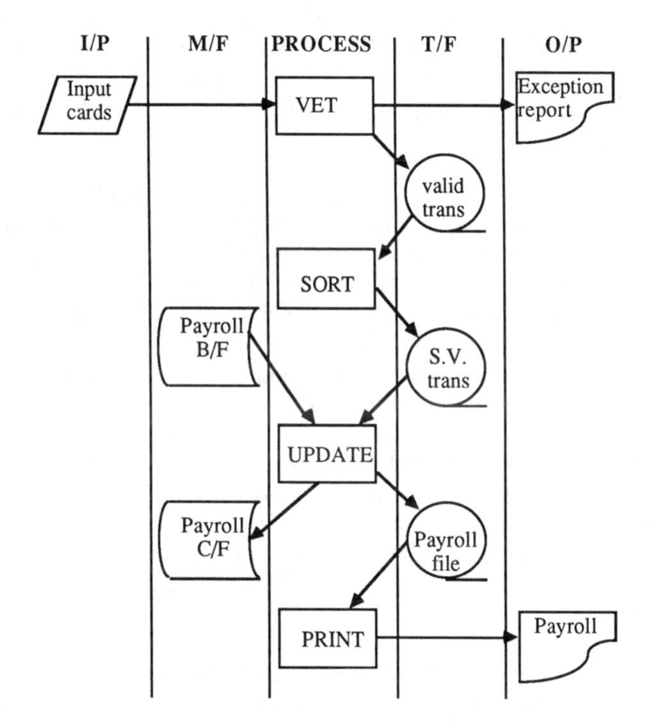

Figure 2.1: A System Flowchart

As mentioned already, traditional techniques concentrate on the description of the solution in computing terms and pay little attention to the understanding and description of the problem domain.

In the following sections, it is proposed that traditional approaches are inappropriate for the development of contemporary information systems and therefore, a more detailed treatment of traditional techniques is outside the scope of this book. Interested readers will find an excellent exposition of traditional techniques in (*Couger et al, 1982*).

2.2 Pressures For Change

The traditional approach to developing information systems is based upon the idea that a problem exists and it can be solved by the use of a computer. An application is considered in terms of its constituent functional parts. The result is the piecemeal computerisation of single applications. Experience however has shown that such an approach cannot be used successfully in complex areas such as the development of integrated systems.

Furthermore, a number of technological changes and user demand for increased functionality of systems are major contributory factors to the necessity for a different approach of developing software systems. The pressures for a move away from the traditional approach may be summarised as follows:

Hardware is Less Important. Until recently an overriding consideration in the development of systems was the hardware that would support the system's operations. Development was often dictated by the cost of the hardware, the result often being a compromise on the users' requirements. With the rapid price decrease witnessed over the past few years, hardware is no longer the focus of attention and therefore the developer is no longer constrained to provide solutions which are guided by hardware considerations.

New Application Requirements. Information system developers are faced with an ever increasing demand for satisfying new application areas such as Office Automation, Decision Support Systems and Expert Systems. In addition, the need for integrated decision-making has highlighted the need for data sharing and integration of information systems.

Shortage of Computer Personnel. The shortage of skilled data processing personnel is well publicised. Coupled to this, there is the problem of the need for specialised development of the existing workforce in order to tackle the ever increasing demand for complex computer-based information systems.

User Dissatisfaction. Problems with the development of information systems usually manifest themselves during operation by users, with the result that these users are more often than not dissatisfied with the delivered product. Late delivery, excessive cost, inflexibility and unreliability are just some of the complaints often voiced by system users.

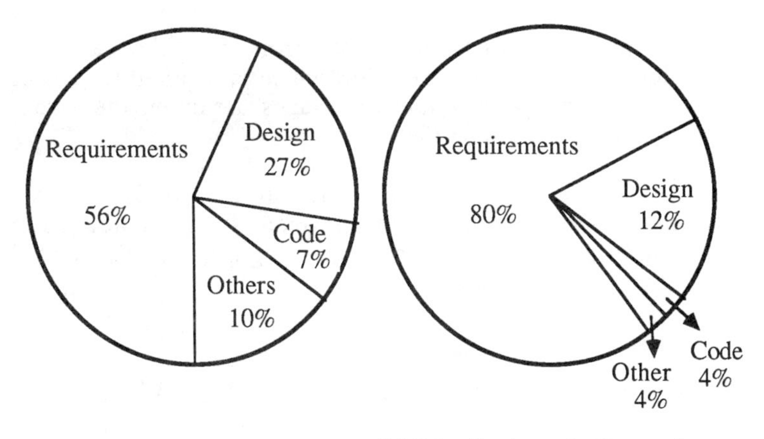

(a) Distribution of errors *(b) Distribution of effort to fix errors*

Figure 2.2: Errors in the Development Lifecycle

During the past two decades system development has suffered from the lack of a coherent approach . This has resulted in the construction of too many systems which fail to fulfil the tasks for which they were commissioned and which become difficult to maintain. As mentioned already, early techniques were based on piecemeal development of separate subsystems without any well defined management strategy. However, this approach has proved to be inadequate because of its failure to handle diverse and complex requirements of contemporary systems. The shortcomings, which are now well documented (*Martin, 1983*) can be summarised as:

- poor requirements specification of systems resulting from ambiguities inherent in narrative descriptions

- poor designs because no rules or heuristics exist for transforming a requirements specification into a design specification

- difficulties in maintaining systems

- progressively more effort is being spent on maintaining poorly designed systems, which means less resource can be devoted to new applications resulting in a hidden backlog of requests from users which has little chance of implementation.

It is argued that 92% of personnel resources is committed to debugging and testing of software systems and this high figure is attributed to the fact that most of the errors occur in poor requirements and design specifications (*Chapin, 1979*; *Talbot, 1985*).

Figure 2.2 (a) shows the distribution of errors over the development lifecycle and figure 2.2 (b) shows the effort which is required for correcting errors over the same set of development stages assuming the errors are detected after the system has gone *live*.

For the reasons outlined in this section, the informal, traditional way of developing information systems has proved to be inadequate for contemporary information systems and have highlighted the need for a proper system development method.

2.3 Information Systems Development Methods

The major response to the problems discussed in the previous section has been the emergence of system development methods. Typical examples of development methods are: SADT (*Ross & Schoman, 1977*), ISAC (*Lundeburg et al, 1981*), SASD (*Yourdon & Constantine, 1977; de Marco, 1978*) and JSD (*Jackson, 1983*). An extensive list of the most widely used methods is given in appendix A.

The characteristics of an information system development method are given below. A method is associated with:

- an underlying philosophy about how the system is to be modelled
- a set of techniques for modelling the system
- a set of tools which are to be used in conjunction with the techniques for the derivation of documentation (from requirements definition models to programming code)
- procedures on how to use the tools
- procedures on planning and controlling the development process
- allocation of people to tasks.

2.3.1 Objectives of a Method

The objective of a software development method is two-fold. Firstly, it attempts to provide the management structures necessary for the planning and control of the development process. And secondly, it attempts to improve the quality of the software produced by providing a formalism for the specification of the system.

The first objective addresses the fact that the requirements and considerations of present-day, large, complex systems are beyond the full understanding by one person (*Lehman, 1978*). This necessitates a team approach to the development task and thus the need for well-defined steps, in order to ensure proper co-ordination and control of teams.

The second objective is concerned with the proper understanding of the system under development, at the appropriate level of abstraction, from requirements specification to code generation and the corresponding documentation which is needed for the effective communication between members of the development team.

There have been several comparative studies of development methods describing the merits of each approach and the parts of system development activity which they address (*c.f. Maddison, 1983; Olle et al, 1982; Olle et al, 1983*). These studies reveal that most of the methods differ considerably, but despite these differences there are a number of common principles which are shared by the majority of the methods. These are:

- a logical model of the system is created and transformed into a physical model by employing a set of design heuristics
- graphical notations are preferred to narrative
- rules for validating the models are explicitly provided
- specification is supported by a set of tools, typically diagrams, text-based specification languages and formalised recording of data in data dictionaries.

In summary, arguably the single most significant factor which differentiates contemporary system development methods from traditional approaches is the fact that these methods place much greater emphasis on the early development stages. Therefore, the main aim of these methods is to minimise the occurrence of errors at the requirements and design stages thus increasing both the productivity of systems personnel and the quality of delivered software.

2.3.2 Method-Related Development Tools

Most contemporary development methods have associated with them computer-assisted tools whose major contribution (currently) is to provide a flexible way of documenting the system under development. In its simplest form such a tool will serve as the *librarian* of the project keeping in an electronic form all the documentation produced by analysts and designers.

Current development tools fall into two broad categories, *graphically-based* tools which support manipulation of diagrams and *text-based* tools. These facilitate storage, editing and sometimes syntax checking of the described data. Generally these tools have evolved in a piecemeal manner supporting single functions in system development, such as data dictionaries. But there is a growing realisation that integrated project support environments are required. These environments contain a wide spectrum of tools which address different development activities, including some of the *management* tasks and clerical tasks (e.g. version control). A further trend is to enhance communication between tools to check the integrity of specifications and designs and to give assistance to developers in an active manner (*Flynn et al, 1986*).

Evolution of support environments can be viewed in terms of generations. The first generation of tools were designed to support single functions in the development process. The most commonly automated facilities were data dictionaries, but tools were also constructed to support specification languages e.g. PSL/PSA and program design e.g. decision table handlers. Most tools were limited in scope and addressed only a single function. Examples of such first generation tools are MAJIC (*Triance & Edwards, 1978*) and PDF, both of which were designed to support the Jackson Structured Programming method. These projects were limited in scope to program design and their products were relevant only to practitioners of a particular method.

The growth of system development methods which placed more emphasis on diagrams and formal notation, created the impetus for a second generation of tools which support the manipulation of diagrams and text in the form of specification languages. These tools generally function independently and support only a single activity in the overall development process. Some verification can be carried out to ensure that the user conforms with the diagramming conventions of the method, but apart from simple validation they are passive, in the sense that they act as a recording instrument for the developer. In the more advanced environments, tools are integrated to a limited extent to provide consistency checking to ensure the objects represented by several tools have consistent identities and characteristics.

As development methods have evolved to increase their coverage of the whole development process, it has become apparent that tools need to be integrated into a systems development environment to support the designer through all stages of specification and design. Such an integration has been achieved in the context of a single method e.g. IEF (*MacDonald, 1986*). Thus only tools specified by a particular design method can be used in one design support environment. Method dependent environments inevitably lock users into a restricted tool set, the size and content of which is dictated by the method's authors and the willingness of software developers to back a method as a market leader.

Current research projects, which aim to create advanced tool environments, have recognised that a variety of support tools need to be hosted by a common interface e.g. PCTE (*Bourginon, 1986*). The research directions have been to produce integrated project support environments (IPSE) and this is one of the aims of both the ESPRIT (*ESPRIT*, 1985) and Alvey (*Alvey, 1983*) Research and Development programmes.

2.4 Classification of Methods

As it can be observed from a simple glance at appendix A, there currently exists a large number of development methods. Each method follows its own approach and has attracted over the years its devotees, as well as critics. The large number of methods presents the student (and teacher) of *systems analysis and development* with considerable problems. Should one follow a particular method or concentrate on general principles? The advantage of the first approach is that one is able to deal in a detailed way with the practicalities of using a development method but has the disadvantage of missing some of the desirable characteristics which are exhibited by other methods. The advantage with the second approach is that the coverage of principles is wider but possibly at the expense of detail and real practical value.

Of course, a practitioner of a method has very little choice but to use the method which is practiced in his organisation and therefore learning about development methods becomes a process of learning about this single method. However, the purpose of this book is not to cover any particular method; if this was the case then a key question would be "which of the hundred or so methods is the most appropriate to cover?" Therefore, the approach adopted in this book is to explore both of the above scenarios. The book explores the principles and techniques found in the *discipline* of system development methods but in order to deal with many pragmatic issues some method-specific aspects must also be covered.

To do this in a meaningful way it is necessary to derive a framework which is shown to represent the backbone of the majority of the development methods and then proceed to demonstrate the concepts and techniques of information system development within this framework. The latter task can be carried out using a notation favoured by particular methods but without loss of generality, since the notation would represent the *externalisation* of concepts shared by many methods (including those that use some other different notation to describe the same concepts).

The required framework is developed here by analysing and classifying the different development methods. A number of research studies have analysed existing methods (*c.f. Griffiths, 1978; Olle et al, 1982; Blank & Krijger, 1982; Maddison, 1983; Martin & McClure, 1984; Floyd, 1986*). Classification of methods in these studies is done in one or more of the following ways: by lifecycle coverage, by lifecycle deliverable, by applicability to application type and finally by philosophy and approach. Of the four classification schemes the most appropriate for a generalised framework is that of the philosophy and approach.

In (*Loucopoulos et al, 1987*; *Black et al, 1987*) other dimensions were found to be useful in classifying methods. Figure 2.3 shows six criteria under which methods may be analysed.

The *philosophy* of a method is a general statement about its basic rationale. In some cases, this will be a particular aspect of a system. In others, it is the strategy used in analysis and design that is the key concept of the method, as for example, in those methods that stress functional decomposition.

The *approach* of a method is a characterisation of its working procedures. In some cases, this is an inflexible sequence of steps, whereas in others, the method is more a set of analytical tools or modelling languages that the user can use as needed, without perceiving a highly prescriptive working procedure.

The *modelling orientation* of a method refers to the modelling mechanisms used by a method to express facts gathered about a system. The external representation is in the form of primitive units aggregated into structures. Such storage schemas can be regarded as the underlying model of a method. The facts contained within a model are then represented to the user in the form of concrete deliverables using a predefined notational style.

The *coverage* of a method is the extent of the development lifecycle that it addresses.

Deliverables refers to the tangible output of a method and, as mentioned above, it is related to the modelling orientation of the method.

Methods (or their authors) may make various *assumptions* for example, about the systems they are dealing with. Such assumptions may be explicitly stated but more often are implicit in the method and its practice.

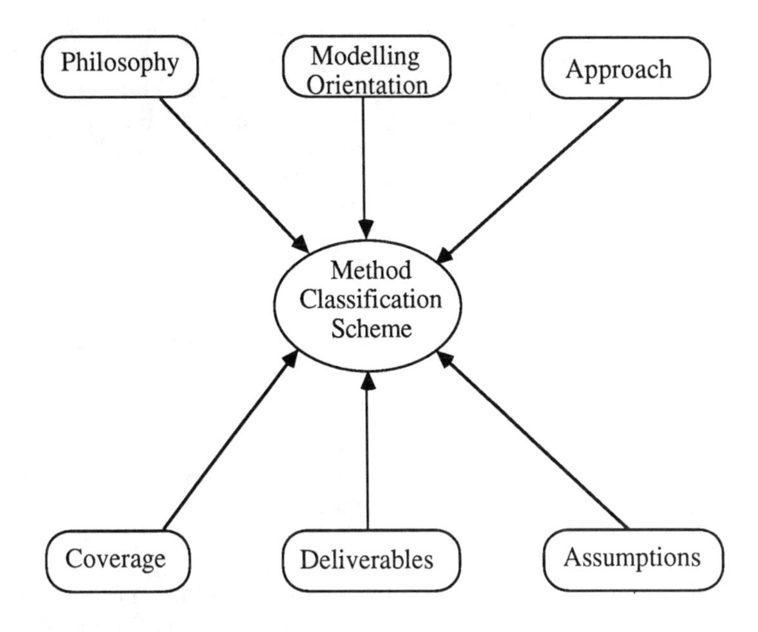

Figure 2.3: Approach to Method Classification

From the above studies, a number of observations can be made about various classes of development methods.

Firstly, there is a core of methods which have a strong and mostly single-purpose philosophical and modelling background. These methods are divided on the emphasis they place on system activities and the objects upon which the system acts. Some methods are functionally orientated (e.g. SASD, ISAC) whereas others (e.g. D2S2, NIAM) place more emphasis on the data of the system. A further variation on this theme is whether time dependencies are emphasised (as in JSD) or not.

Secondly, a different class of method (e.g. SSADM, Information Engineering) attempt to increase their coverage by increasing their impact on the development lifecycle. This, they attempt to achieve by widening their modelling horizons to encompass a number of modelling orientations which would normally require the use of more than one method from the first category.

Thirdly, there exists a class of methods which can be regarded as a new generation of methods with their emphasis being on either formality of system specification (e.g. VDM, HOS) or semantic richness (e.g. ACM/PCM,

TAXIS). The first category aims to prove correct system behaviour by reducing system activity to a formal mathematical notation based on a variety of techniques including set theory, relational algebra and calculus. The second category uses ideas derived from the field of artificial intelligence and uses knowledge representation schemes based on semantic networks, frames and first order logic. This category owes its heritage to a convergence of conceptual database modelling and knowledge representation work (*Chen 1976; Brodie & Silva 1982; Greenspan & Mylopoulos 1982*).

2.5 A Framework for Analysis and Development

In section 1.3 it was argued that development of an information system is concerned with the development of models and that these models pertain to a system's *dynamic* and *static* aspects and its *interaction* with its environment. This view is very useful in identifying one dimension of the required framework.

However, this in itself is not sufficient. There is a need to consider at least a second dimension which is concerned with the orientation of the methods. In considering the methods' philosophy, coverage, modelling background and approach, four main areas can be identified. These are: *problem analysis*, *functional analysis*, *conceptual data modelling (data analysis)* and *system design*. The first three are concerned with the expansion stage whereas the fourth relates to the contraction stage of the development process.

Figure 2.4 shows the framework which relates to these two dimensions and which is adopted in this book for the purpose of discussing the concepts and techniques of *systems analysis and development.*

The *problem analysis*, *functional analysis* and *conceptual data modelling* areas deal with a single fundamental aspect of information systems, interaction with the environment, processes and data respectively. The *system design* area, being concerned with a *synthesis* of components into a new information system, deals of course with all three fundamental system aspects.

Each cell of the matrix refers to the chapter which deals with the corresponding material.

	DYNAMIC	STATIC	INTERFACE
FUNCTIONAL ANALYSIS	PROCESS (Chapter 4)		
CONCEPTUAL DATA MODELLING		DATA (Chapter 5)	
PROBLEM ANALYSIS			INTERACTION (Chapter 3)
SYSTEM DESIGN	PROCESS (Chapter 7)	DATA (Chapter 8)	INTERACTION (Chapter 9)

Figure 2.4: A Framework for Analysis and Development

2.5.1 Problem Analysis

Problem analysis is concerned with the examination of the environment within which the future information system will operate. The process serves as the means of understanding the user environment and the tasks which are carried out within such an environment. At the outset it will be the means of identifying the key problem areas and the major sources of facts which an analyst must collect in order to build a model of the functional and non-functional requirements of the information system.

Allied to the need for the correct identification of the problem areas is the need for consultation with the potential users of the system and their effective participation in development process.

User participation has been advocated by a number of researchers in Computer as well as Management Science fields and is the basis of the *socio-technical* approach (*Taylor, 1975*). The participative approach is defined at three levels giving rise to three forms, namely those of *consultative participation*, *representative participation* and *concensus participation* (*Mumford et al, 1978*).

2.5.2 Functional Analysis

The term *functional analysis* is used in this book to refer to a set of techniques, including the so-called structured techniques, which are primarily concerned with the specification of the functions of an organisation and the activities (or processes) which take place within these functions.

These techniques pay particular attention to the description of the problem and at requirements and design level they hardly acknowledge the existence of a computer. The philosophy of functional analysis is based upon the following:

- enabling developers to partition complex or large systems into smaller, more manageable units
- encouraging the separation of the logical view of a system from its physical characteristics
- adopting the use of graphics for easier communication
- encouraging detailed system documentation.

Functional analysis techniques use mostly graphical notation to represent a specification. The notation used has been termed the *box and arrow* approach and, although there are some slight variations from one method to another, most notations are concerned with a limited set of primitive constructs from which orderly construction of any size specification can be composed.

Chapter 4 demonstrates the concepts and techniques of functional analysis using the notation found in SASD (*Yourdon, 1977*) but it should be emphasised that the principles covered in that chapter are found in many other methods which also deal in functional analysis, methods such as STRADIS (*Gane & Sarson, 1979*), SADT (*Ross & Schoman, 1977*), ISAC (*Lundeberg et al, 1981*) and many others. These concepts are also found in methods which are not considered as being principally *functionally oriented*, such as Information Engineering (*MacDonald, 1986*) and JSD (*Jackson, 1983*).

2.5.3 Conceptual Data Modelling

Following the recommendations of the ANSI-SPARC three level architecture (*ANSI, 1975*), a number of modelling schemes were put forward which concentrated on the data of an organisation. A number of researchers have highlighted the importance of conceptual data modelling to database design

(*Nijssen, 1977*). Nowadays conceptual data modelling is used not only in designing databases but also, in a wider context, in the requirements and design specification stages of information system development (*Chen, 1976*; *Codd, 1979*; *Greenspan & Mylopoulos, 1982*; *Borgida et al, 1982*; *Albano et al, 1985*).

Data modelling is based on the idea that data is the fundamental building block of a system. Data modelling encourages the specification of a logical model of all organisational data which serves as a single reference point for all applications. The specification is expressed in such a way so as to conform to the ISO *conceptualisation principle* (*van Griethuysen et al, 1982*). This states that: "A conceptual schema should only include conceptually relevant aspects of the universe of discourse, thus excluding all aspects of (external or internal) data representations, physical data organisation and access as well as all aspects of particular external user representation such as message formats, data structures etc".

2.5.4 System Design

System design is concerned with synthesizing the three types of model (interface, process and data) into a design specification which will serve as the basis of implementing the desired information system.

The techniques involved aim at establishing a strong framework upon which programming will be based and make use of the principles of decomposition and modularity. The techniques aim to produce a solution which is structured in a way that matches the structure of the problem. The objective is to specify a design in such a way that will result in data processing systems could be successfully modified and enhanced in the future.

Contemporary development methods follow one of two approaches to system design, either data-oriented or process-oriented.

Data-oriented approaches, for example JSP (*Jackson, 1975*) aim to design the system according to the behaviour of its data. On the other hand process-oriented approaches, for example Structured Design (*Yourdon & Constantine, 1977*), concentrate on system modules, their interrelationships and in the application of heuristics for the coupling and cohesion of the modules. Both of these approaches are described in chapter 7.

2.6 Characteristics of Systems Analysis and Design

The remaining chapters of this book are devoted to the concepts and techniques for analysing and designing information systems. However, it is worth first examining the characteristics of these tasks.

The tasks of analysis and design of information systems are complex and many argue that the continuing problems with the development of such systems may be attributed to the nature of the work. The activities involved, particularly at the early phases of development, are non-algorithmic, are long and iterative and involve considerable informality and uncertainty. The task of systems analysis has been given the following characteristics (*Vitalari & Dickson, 1983*):

- because the task is carried out at the early stages of the development lifecycle it is often difficult to define the boundaries of the universe of discourse in a deterministic way

- because there is little structure of the problem domain before hand there is a considerable degree of uncertainty about the nature and make-up of the possible solutions

- analysis problems are dynamic that is, they change while they are being solved because of their organisational context and the multiple participants involved in the definition and specification process

- solutions to analysis problems require interdisciplinary knowledge and skill

- the process of analysis itself, is primarily cognitive in nature, requiring the analyst to structure an abstract problem, process diverse information, and develop a logical and internally consistent set of models.

A number of empirical studies have been carried out in an attempt to better understand the process of developing the tasks of analysis and design. The differences in behaviour between high and low-rated developers were investigated in separate projects (*Vitalari & Dickson, 1983*; *Fickas, 1987*; *Adelson & Soloway, 1985; Curtis et al, 1988*).

Based upon the results of these studies, it is possible to identify several major, although not mutually exclusive, types of working practices by system developers. These practices may be summarised as follows.

Developers Use Analogy. Developers use information from the environment to classify problems and relate them to previous experience. Empirical studies have shown that experienced developers begin by establishing a set of context questions and then proceed by considering alternatives. Much of the contextual information depends on previous knowledge about the application domain and the analogies that a developer will establish on the basis of such knowledge.

Developers Build Hierarchies of Models. Expert developers tend to start solving a problem by forming a mental model of the problem at an abstract level. This model is then refined, by a progression of transformations, into a concrete model, as more information is obtained. Developers are aware of the various levels of policy within a domain and use this knowledge to guide a user during a requirements capture session.

Developers Formulate Hypotheses. Hypotheses are developed as to the nature of a solution, as information is collected. An experienced developer not only uses hypothetical examples to capture more facts from a user but also to clarify some previously acquired information about the object system. Experience in the application domain seems to be an important factor in formulating likely outcomes of the solution space.

Developers Use Summarisation. Developers almost always summarise in order to verify their findings. It has been observed that during a typical user-analyst session the analyst will summarise two or three times and each time the summarisation will trigger a new set of questions (*Fickas, 1987*). Summarisation may in order to clarify certain points from previous discussions or to encourage participants to contribute more information about the modelled domain.

From this analysis it can be argued that the success or otherwise of a system depends to a large degree on the experience, knowledge and general competence of the persons that carry out this activity. Furthermore, the process itself is very informal and thus it is not surprising that in practice this leads to informal specifications (normally in narrative) with serious consequences.

Therefore, it is of crucial importance that software development receives attention with the objective to overcoming the problems inherent in the informality of the process. A methodical approach which uses appropriate concepts, techniques and tools seems to go a long way towards meeting the objectives of developing *effective* and *flexible* systems.

2.7 Summary

Traditional approaches to developing information systems involve the piecemeal computerisation of individual application areas. The traditional approach has proved to be inappropriate for the development of large and complex contemporary systems and therefore new approaches have been advocated. New development techniques address the problems of firstly analysing system requirements and the structural relationships of these requirements, secondly designing the system so that its structure matches the structure of the problem and finally implementing the design using simple logical structures.

Chapter 3

Problem Analysis and Fact Collection

In recent years, experience in software development has shown that if insufficient time is spent on each stage of software development, the cost of correcting errors detected after the system has been delivered increases exponentially. It therefore follows that in order to minimise software development costs, a software developer must have a clear and comprehensive understanding of the desired system and its requirements before large-scale development begins; and it is the role of the systems analyst to establish this understanding and embody it in a precise and unambiguous specification.

This chapter begins by outlining the initial steps of analysis, followed by a review of the basic fact collection techniques employed by systems analysts.

3.1 Overview of Problem Analysis

Problem Analysis is about analysts seeing and understanding the world through other peoples' eyes: managers, system users, customers, technicians and so on. The starting point of this analysis can vary considerably, from volumes of documentation arising from a feasibility study, through to a brief statement of requirements, such as "We need a system to do..." and successful system development requires the analyst to investigate these statements and acquire the expertise, knowledge and understanding of people within the organisation under study. This is achieved by gathering, sifting and analysing information provided by individuals and building it into a clear description of the problems within an organisation.

Figure 3.1 shows a symbolic view of what an analyst is seeking to achieve; that is to view the user's problem domain and establish its boundaries. Notice that the analyst does not necessarily have to have the same perspective or angle of view as a user, but his field of vision and understanding of what he sees must coincide with that of the user.

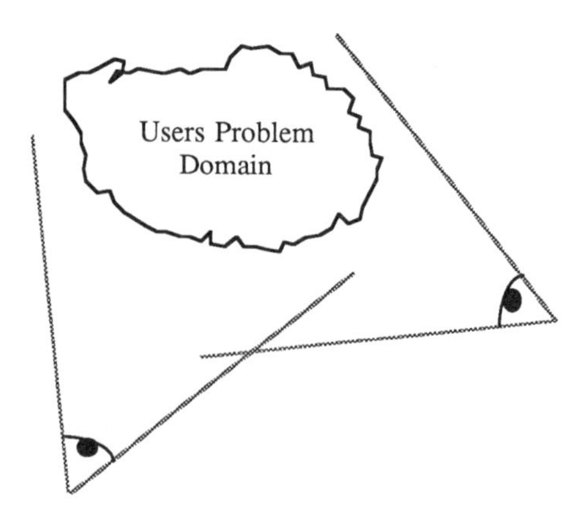

Figure 3.1: A User's and Analyst's view

Unfortunately, achieving such an understanding is not easy since an analyst must deal with people, perceptions and practices which, unlike mechanical devices, can be unpredictable, irrational and unreliable. Typically the problems faced by an analyst will include the following.

Locality of View. Experience shows that the human mind has a tendency to concentrate on the fine detail of a problem or system at the cost of neglecting the general or higher-level view. As a result, during the initial stages of analysis, in which the analyst is trying to build a wide picture of an organisation, users tend to provide analysts with information concentrating on detail and procedure, rather than the overall objectives and purpose of their particular function. A major goal of analysis is therefore to abstract from this detail and derive the general principles and objectives of a system.

Inconsistency of View. One of the biggest barrier to effective analysis is the wide, apparent inconsistency that occurs between individual views of an organisation and its systems. For example, one part of an organisation, such as a sales department might regard a particular document as a customer order, whilst the same document may be regarded by a purchasing department as either a despatch note for goods that are currently in stock or as a purchase request for those goods which are currently out of stock. Such apparent inconsistency of view arises simply because of the different ways in which individuals or organisational units process information. In the above example, once the analyst has identified that the same document is being referenced, a standard name can be adopted. A more problematic inconsistency of view is where individuals genuinely misunderstand the meaning of information or the objectives of some process. Correction of this situation requires greater investigation and a reconciliation of views, with individuals needing to rethink or change their perceptions.

Incompleteness of View. Another major problem faced by analysts is the problem of knowing how complete their view is of a system and identifying when all possible user requirements have been established. In practice, this can only be overcome by constant user validation and ultimately is terminated by a system description which can adequately meet the users' needs. Notice the implication that such descriptions are not necessarily optimal.

The Perceived Threat. Many employees, most of whom are computer illiterate and unaware of the capabilities and limitations of computers, perceive the computer as a threat to their status, working practices and future employment (*Mumford et al, 1978*).

It is for reasons such as these that systems analysis and investigation is regarded as the hardest part of the software development process and yet by the same virtue makes it the most interesting task within that process.

So what is an analyst trying achieve? In essence, it is an understanding of what an organisation and its users require, often when they do not fully know themselves. It is therefore the role of the analyst to probe, explore and document the organisation's business, trying to understand its objectives, how it operates and to identify systems which will enable the organisation to run more efficiently. The beginning of this process is concerned with understanding the organisation, its problems and needs and typically an analyst will start to build such a picture through the following five, basic steps:

- determining the organisational structure and its purpose
- determining the problem areas within the organisation
- determining the environment around the organisation and identifying the boundaries of its problem
- establishing a detailed specification of the problem
- validating the information gathered.

There are several techniques used to establish the information necessary to perform these tasks; they are documentation review, interviewing, questioning, observation and measuring.

Whilst these techniques and steps are presented in a sequential fashion, it must be emphasised that in practice they are not discrete, ordered steps, at the end of which an analyst has a clear view of the purpose of a system. Instead each step will be undertaken in parallel with each other, with the analyst developing an evolving view of the required system. Similarly, a good analyst will not leave validation until the end of the problem analysis process. Instead, validation of gathered facts and opinions is a task which should occur at regular intervals, for example, following an interview or after a questionnaire has been circulated, completed and returned.

The remainder of this chapter reviews in detail the main problem analysis steps and fact collection techniques.

3.2 Steps in Problem Analysis

3.2.1 Determine Organisational Structure and Purpose

The start of a system development exercise is normally triggered by a request to examine an activity or function within an organisation, with the purpose of diagnosing some problem. Subsequently this will lead to a set of possible solutions, possibly computer-based, from which a desired solution can be selected. Until the eventual design and implementation of a system, all solutions should be considered and the decision between a manual or computer-based solution should not be prejudged. There are many systems which can be implemented without computer technology and it is important to consider the cost-benefits of the various solutions available.

The first activity in this exercise is to determine the structure of the organisation under study and gain an appreciation of its objectives and activities. Knowledge of this kind is generally known as *domain knowledge* and although it may directly contribute little to establishing the overall problem, it is nevertheless vital that analysts establish such knowledge in order to build a broad understanding of the organisation and its activities. This may be likened to a lorry driver who is to deliver goods to a particular road within a certain town. Sometimes, the driver may know the area, in which case there is little difficulty. However, other areas may not be so familiar to the driver and so he would probably buy a road map or ask a colleague to describe the area.

In the same way, when analysts are called into a particular organisation, sometimes they may know the organisational structure and purpose, but in the majority of the cases they probably will not. Therefore, the first job would normally be to build a model of the organisation which can subsequently be used as a *map* of the organisation. Figure 3.2 shows a sample organisation structure for the hotel case study. It can be seen that the hotel consists of three main areas: domestic services, administration and catering, each of which are further decomposed.

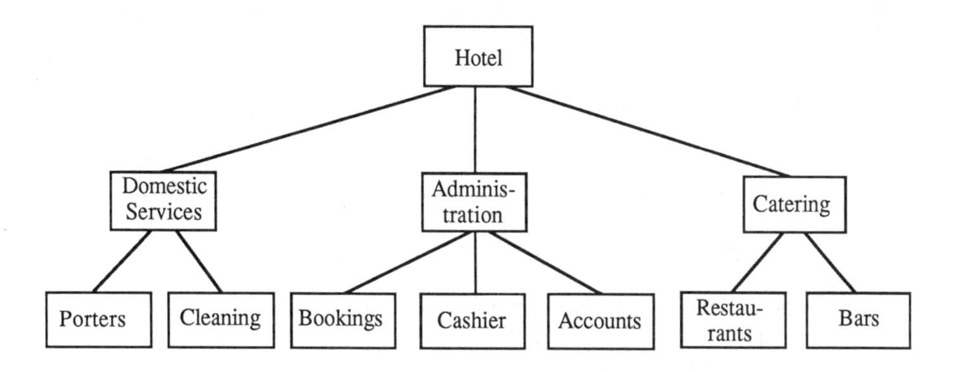

Figure 3.2: A Sample Hotel Organisational Structure

3.2.2 Determine the Problem Area

Once an understanding of the structure of an organisation has been developed, the analyst can begin to determine the true nature of the problems its faces. The word *true* is used purposefully, not to imply deceit upon the part of those who may commission a study, but because the initial statement of an organisation's problem is often vague, ambiguous or merely describes the symptoms of a deeper, underlying problem.

For example, a wholesale distribution company may be aware that it is loosing sales orders because "customers complain of a long time between placing an order and receiving the goods". At first sight, one may be tempted to investigate the movement of customer order requests from the sales department to the warehouse or even examine the efficiency of the transport department. However, the truth of the situation may be that the purchasing department has an inaccurate knowledge of its stock levels and is only reordering stock when it attempts to satisfy a customer's order and finds that there is none left.

Therefore the analyst, through the techniques of fact collection (which are discussed in section 3.3) must identify the area of an organisation's activities in which problems are occurring and on the basis of this, produce a simple statement of the problem. In the stock example, the problem may be summarised as "the need to provide a more reliable stock level report so that the purchasing department can provide a continuous supply of stock".

3.2.3 Determine Environment and Boundaries

Once a general problem area has been identified, the analyst must next describe and understand the environment in which the organisation operates. This is necessary so that a clear perspective of the organisation can be developed, together with an understanding of how the particular problem area interacts with those people or other companies and agencies immediately outside the organisation.

An environment can be regarded as consisting of the following factors:

- management and administrative objectives
- the characteristics of the business or activity in which the organisation is involved
- the availability of resources for use by the organisation
- legal requirements, statutes and governmental regulation
- public opinion.

In the case of the hotel, it is likely that the management objectives of the organisation is to make an adequate return on capital investment through the maximisation of room occupancy in its hotel and consequent receipt of sales money in a greater volume than the cost of the operation.

Typical of how environmental characteristics may affect a business and its system requirements would be the profile of room demand and clientele, in the case of a hotel system. For example, a hotel in a busy city centre, catering largely for business travellers staying mid-week would have different requirements from a room booking and invoicing system to a seaside hotel catering mainly for summer holidays. Figure 3.3 presents a selection of their respective characteristics. From these characteristics, it can be seen that the city centre hotel would need a fast, efficient check-in and check-out system, with automated charging for bars and restaurants and subsequent invoicing on check-out. The seaside hotel, on the other hand, may place greater requirements on support for book-keeping functions, leaving check-in and similar facilities to a manual system. Of course, only with further elaboration by users can the analyst fully determine the system requirements, but the example does show how supposedly the same situation may lead to different requirements.

All environmental factors need to be fully understood since they will impact the requirements of a system and will effect decisions taken during system design, especially when several alternatives exist. Such factors will further determine the nature of a system through effecting:

- the availability of data within the organisation
- the rate of growth of the organisation's activities
- the manner in which a computer-based system may have to operate.

City Centre Hotel	Seaside Hotel
Mid-week trade peak	Summer trade
Conference trade at Easter	Weekly bookings
Christmas parties	Weekend arrivals
Single and double rooms	Small cash bar
Late checking in	Separate restuarant
Early checking out	
Large restaurant turnover	

Figure 3.3: Contrasting Hotel Characteristics

Once environmental factors have been fully understood, the analyst needs to place bounds on the activities of the organisation and clearly identify the interfaces across such boundaries. For example, if analysts are to restrict themselves to the invoicing activities within an organisation, they must know what information flows in and out of the cashiers section. This will help them to ensure that all necessary information is output by the system and identify any new information needs of the system.

Once the analyst has a picture of an organisation, the information gathered so far can be formalised in a business system model which simply describes the inputs and outputs to an organisation, the operations conducted by the organisation and the resources available to the organisation. Of course, in a large organisation, such a model may refer to only a part of the organisation.

An example of a business system model is shown in figure 3.4. It can be seen that the organisation uses its room, restaurants, bars, staff etc., to accept conference bookings, casual trade and banquets and through its catering, room letting, cleaning and invoicing operations, achieves sales, bookings and provision of information.

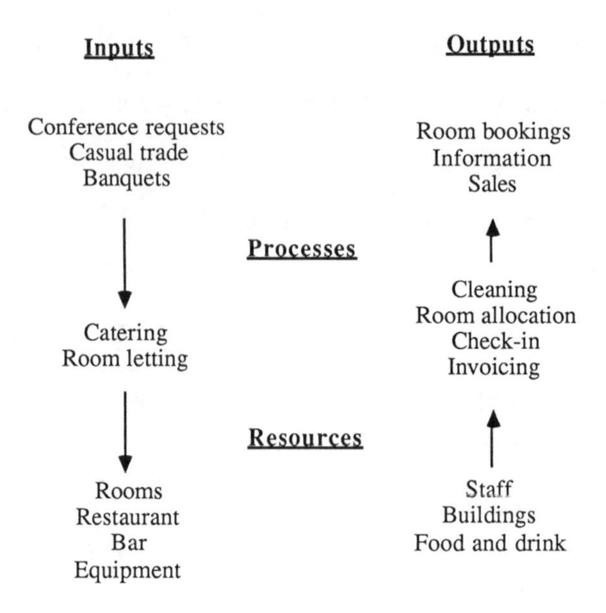

Figure 3.4: A Business System Model

3.2.4 Establish Detailed Specification of the Problem

Having established a broad picture of the functioning of an organisation and the problems that pertain to it, the analyst must next obtain specific details of the key components within the problem boundary. Generally, such detail would include:

- identification and brief description of key personnel and processes such as check-in desk clerk, room service personnel

- identification and brief description of the main items of data used such as room booking, charge bill, refund note

- details of access to the data including the frequency of access and volumes of data, such as when are the peak periods for hotel check-in and check-out and how transactions are made per hour during the busiest period

- policies and practices relating to the organisation's activities, such as the credit rating of customers or governmental regulations dictating the maximum number of people resident in a hotel overnight

- exceptions to the norm, such as special procedures or policies, abnormal loads, work of a cyclical nature and techniques for error handling, such as identification of peak conference seasons in which the hotel is operating to capacity

- legal constraints such as union agreements, company rules or government legislation.

The majority of this information will be unstructured during the initial stages of system development, but as the analysis phase progresses, the various items of detail will fit together in a structured and coordinated fashion until they are built up into a unified model of the entire system.

3.2.5 Validate Information Gathered

Once information has been gathered and analysed, it is important that it is reviewed and validated by users in order to ensure that the analyst has not misunderstood or misrepresented the views, opinions, facts and measurements that have been obtained.

A variety of techniques exist for this activity, such as reviews, walkthroughs and inspections, but as they apply to all stages of analysis, their discussion is left until chapter 6.

3.3 Techniques for Fact Collection

3.3.1 Overview

So far, the various steps for determining an organisation's problem have been identified. However, nothing has yet been said about how particular items of information are gathered. Fortunately, the techniques used for achieving the different steps are usually the same and consist of combinations of:

- documentation review

- interviewing

- questioning

- observation

- measuring.

All of these techniques involve, not only the analyst, but also a wide range of staff within the organisation under analysis, some in a passive, observing role and others in an active and contributing role. The process is thus heavily oriented towards personal contact and relationships and because of this, most fact collection techniques are generally difficult to do well.

3.3.2 Documentation Review

Documentation of existing systems is often a useful starting point in the early stages of analysis since many system manuals present an overview of the current system, together with summary information. Furthermore, documentation generally has the advantage of being portable and easily accessible and thus can act as a useful springboard for further investigation and fact collection. However, documentation also suffers from a number of problems.

Firstly, in many large systems the volume of documentation is often excessively large and most adds adds little useful information to the analyst's view of a system. The analyst therefore needs a clear summary and index to the documentation if the most effective use of the material is to be made.

Secondly, much documentation is very sketchy in nature. This partly arises because many system developers regard documentation as the job to do last and much of the detail of a system is consequently forgotten and omitted.

Finally, documentation is notoriously inaccurate and out-of-date since it is widely believed that it is better to ensure that the actual software works rather than to document how it should work.

Consequently, documentation should only be relied upon for a very general picture of an aspect of an organisation's activities. A checklist of potentially useful documents include the following.

- Instruction manuals detailing company organisation, policy and existing systems: such manuals are normally only available in the largest organisations and usually fail to reflect the most recent modifications and amendments.

- Job descriptions and staff duties: such documents should always be confirmed with individual employees as they are prone to frequent change as employment duties evolve.

- Previous reports and investigations: preferably the most recent and with a check to ensure their relevance to present day activities and confirmation that any assumptions made still apply.

- Publicity and promotional material: such material is generally not very technical, but it can provide a useful overview.

3.3.3 Interviewing

Interviewing is probably the most important tool used by an analyst and it is generally accepted that effective analysis can only be carried out when a degree of interviewing is undertaken. However, the technique can only effectively be developed through practice as its success relies not only upon proper preparation, but also on the way in which an analyst conducts interviews. The particular skill necessary is the ability to ensure, on the one hand, that the interviewee keeps to the subject, whilst on the other hand the interviewee is given sufficient freedom to introduce points that may be of relevance to the analysis.

It is extremely important that if the best use of an interview is to be made, analysts prepare thoroughly beforehand. This preparation must include the following.

- A definition of the purpose of the interview: what information is missing in the analyst's picture of an activity or organisation and what questions must be asked to complete this view.

- A list of who must be interviewed; there is no point in conducting an interview with an office junior about the overall aims and objectives of an organisation for example. Furthermore, once potential interviewees have been identified, approval for their participation from their manager or possibly the interviewees themselves must be sought.

  Figure 3.5 provides some suggestions as to the relationship between organisation function and the types of information known by individuals within that function.

- A degree of homework which includes a full understanding of technical terms, a review of appropriate documentation and facts and figures, the preparation of questions to be asked during the interview.

During the actual interview, the analyst should attempt to conduct the proceedings in three basic phases: the opening phase, which outlines the objectives of the interview; the main fact collecting phase and the concluding phase, which summarises the main points of discussion.

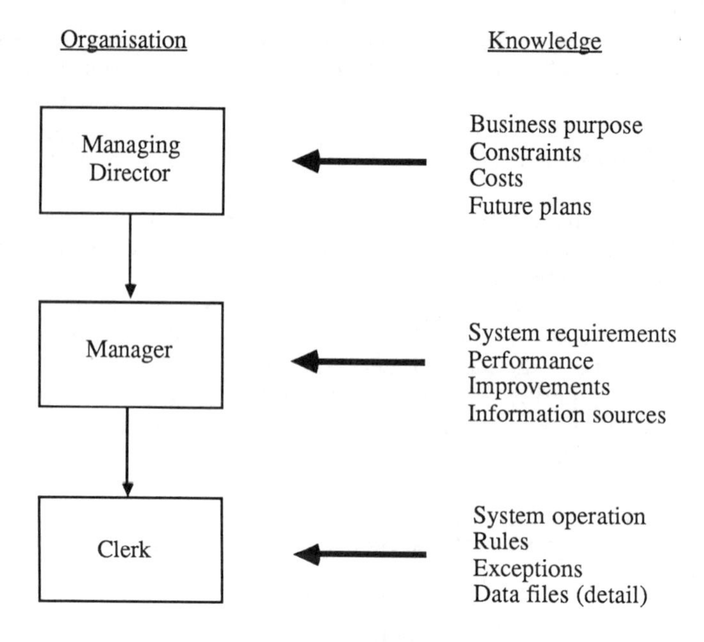

Figure 3.5: Levels of Knowledge

The objectives of the opening phase of the interview is to develop a rapport between the analyst and interviewee. Topics covered in this phase should include the following.

- An explanation as to the objectives of the interview, so that the interviewee is fully aware of the boundaries within which the analyst will want to ask questions and, in the case of a thoughtful and co-operating interviewee, answers can be restricted to the area of interest to the analyst.

- An explanation as to the background of the analysis; what the aims of the overall study are; how it might affect the interviewee and the benefits of any potential outcome of the study.

- The reasons why the interviewee was chosen, identifying any skills or experience the interviewee was believed to hold.

- A degree of smalltalk that will relax the interviewee.

The main part of the interview should consist of the analyst posing a series of questions on different aspects of the organisation's activities, followed by supplemental questions to clarify or extend the points raised during the interviewee's answer. The main points that the analyst should bear in mind are as follows.

- Start with a general, opening question which introduces the main subject area.

- Follow the opening question and initial answer by a series of supplemental questions on specific detail, policy or actions, each question becoming more specific.

- Listen to the answers, making notes where relevant and ensuring that the interviewee keeps to the point.

- At regular intervals, the analyst should attempt to summarise the facts stated by the interviewee and invite them to confirm or amend this summary.

Figure 3.6 summarises this iterative activity of question initiation, reaction and clarification.

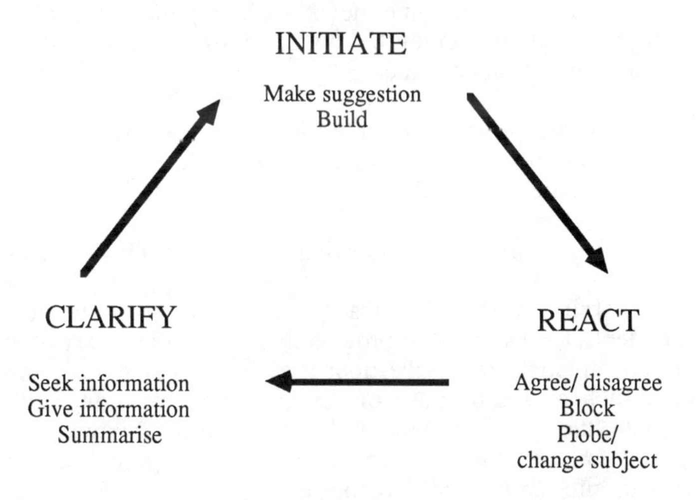

Figure 3.6: Iteration in the Interviewing Process

Once the main body of detail has been gathered, the analyst can begin the process of drawing the interview to a close. During this stage, the main points raised in the interview should again be summarised and confirmed, together with any outstanding questions or loose-ends tied up.

When the interview is complete, it is important for the analyst to review the notes taken and to supplement this wherever they appear to be sketchy. Furthermore, like all the remaining fact collection techniques, statements of fact should be checked and verified using alternative sources wherever possible.

3.3.4 Questioning

Another popular fact collection technique is questioning, in which a respondent completes a questionnaire, an example of which is shown in figure 3.7. However, it is a technique that is difficult to undertake well and requires extremely careful question design to avoid problems of misunderstanding, vagueness and ambiguity.

Like many of the other fact collection techniques, questionnaires should be regarded as just one of a range of techniques that can be employed during analysis, each technique supplementing the other. In this context, a questionnaire can be used for two purposes:

- as a follow-up to an interview, in which points of detail, facts and figures which cannot be obtained or remembered during an interview are documented

- as a filter prior to an interview or observation exercise, the information obtained being used to highlight particular areas and points to be addressed.

In designing a questionnaire, the following issues should be borne in mind.

Covering Instructions. When a questionnaire is to be left with a respondent, it is essential to provide them with a clear set of guidelines on the completion of the questionnaire. These guidelines should cover points such as the purpose of the questionnaire, how and when to complete the questionnaire and the authority by which they are required to complete the questionnaire. The guidelines should be personal, informative and easy to use.

Question Content. When posing any question, the analyst must ensure that it encourages the respondent to give an answer that reflects the true state of affairs. A big problem with questioning, either by interview or questionnaire, is that respondents often feel that they must give an answer to a question whether or not they are accurately knowledgeable about the subject area. Thus a common technique employed in question content design is to ask opinions, rather than for hard facts, unless of course one is reasonably sure that the respondent has the necessary information to hand.

In the sample questionnaire in figure 3.7, it would obviously only be appropriate to give the questionnaire to staff working on the *front desk* of a hotel.

HOTEL ADMINISTRATION SYSTEM
Questionnaire

The purpose of the questionnaire is to discover some of the problems with the current hotel administration system, in preparation for the development of a new system. Your views and opinions will be greatly valued.

1. Please state your name:____________ position:______________

2. How long have you been employed in the hotel?

   ☐ less than 6 months ☐ more than 6 months

3. What do you consider to be the main problems during check-in and check-out of a guest?______________________________
   ____

4. Please indicate, using a scale of 1 to 5, the busiest check-in and check-out periods. 1= the busiest, 5= not very busy at all.

	Check-in	Check-out
before 7.15am		
7.15-7.30am		
7.30-7.45am		
7.45-8.15am		
8.15-8.45am		
8.45-10.30am		
10.30am-4pm		
4pm-5pm		
5pm-6pm		
6pm-7pm		
after 7pm		

5. What are the main enquiries for information that you receive from guests?______________________________
   ____

Please turn over...

Figure 3.7: A Sample Questionnaire

Wording. A number of general principles apply to question wording and these include the following.

- Simplicity: technical jargon, especially in computing, often means many things to many people and should be avoided, unless the analyst can be certain that a particular term is fully understood both by himself and potential respondents. A useful technique here is to construct a glossary of organisation-specific terms during an analysis.

- Conciseness: one word should be used instead of many since this will make the question easier to read and to understand.

- Vagueness: when a specific answer is wanted, a precise and specific question must be asked. For example, to ask what proportion of one's time is spent on a particular activity, may produce the reply, "most of my time", which says virtually nothing! The correct question should be something like, how many hours per week do you spent on activity X? Similarly in the sample questionnaire in figure 3.7, the analyst has not simply asked, "when are the busiest check-in and check-out times, but rather has attempted to get a detailed breakdown.

- Ambiguity: ambiguous questions, as opposed to questions using technical or complex phrases, are those such as "is your productivity made more difficult because of poor communications?" In this question, it is not clear whether the analyst is interested in poor communications or productivity; potential cause is linked with effect. It would be better to first ask about productivity, followed by a separate question on communications.

- Memory: questions involving memory are often a source of inaccuracy in questioning, the effects of which may be the exaggerated importance or insignificance of an event or memory loss leading to guessed answers. In these cases, as with all fact collection techniques, it is important to build cross-checking questions into a questionnaire or use other fact collection techniques to verify answers and thus ensure consistency and accuracy.

- Ability to answer: when asking a respondent to answer a question, it is vital that they have the knowledge to answer. For example, in the sample questionnaire, users are asked to *rate* how busy they are at certain periods, rather than give the actual average number of check-ins and check-out they perform. To provide such detailed information would probably be virtually impossible.

Open and Pre-Coded Questions. When questions of opinion are concerned, an open-ended question technique is best in which the respondent is free to write a sentence or two expressing his views. For example, question 5 in the sample questionnaire is open-ended because it is seeking to capture knowledge about possible system problems that the analyst who produced the questionnaire does not necessarily have.

On the other hand, questions of fact are more suited to a pre-coded technique in which all potential answers or ranges of answers are listed and the respondent selects one or more. In this latter case, the analyst must ensure that all possible answers are listed and this can usually be achieved through a pilot survey or some other initial fact collection exercise. In question 4 in the hotel survey, it is likely that the analyst has conducted some prior investigation to define the main periods of the day for checking-in and checking-out.

Question Order. The issue of question order is important in that the respondents confidence must be built up during the answering of the questionnaire. The following guidelines should therefore be followed.

- The easiest questions should come first, thus allowing a rapport to grow between the respondent and questionnaire

- The questions should follow in a logical manner, not requiring the respondent to make *mental leaps* between different issues, but also ensuring that the flow of the questions does not suggest answers to respondents or lead them to write down answers which they believe the analyst wants or expects.

3.3.5 Observation

Observation entails watching and monitoring the activities conducted by employees within a particular area of an organisation which fulfil the aims of that organisation.

The advantages of observation is that it is a useful technique when other sources of information are not available, for example, when there are no previous records of an activity or the answer to questions in an interview or questionnaire are unknown, vague or require verification. Furthermore, observation can remove the subjective reactions of people who are questioned in an interview in which they feel they must give some answer, even if they know it to be inaccurate, so as not to loose face. Other advantages of observation include the following.

- Observation provides the ability to identify and gain a sensitivity to abnormalities, interruptions and the peaks and troughs in an individual's workload which cannot normally be detected through statistics such as the number of hours an individual spends on a particular activity in a week.

- Observation helps to identify the informal patterns of communication and behaviour within an organisation.

- Observation also helps to establish the patterns of file and document access.

However, because of the high degree of concentration required, together with being a very time-consuming activity, observation is generally avoided where possible.

Inherent in this method of fact collection are a number of problems, arising from the psychological effects of one person, the analyst, watching somebody else, an employee, executing their role. Behaviour modification by an employee will always be prevalent: for a variety of reasons, the employee will be self-conscious, work slower or work harder, thus obscuring the view of the analyst.

Other problems with observation include the following.

- Observation is of no use for activities which have ceased, nor for activities which take place at remote sites or at unusual times of the day.

- Observation is an inappropriate technique for the study of beliefs and attitudes.

- *Overhearing* often gives an unrepresentative view of a situation caused by the analysis of a comment or behaviour outside a wider context of the general background and environment of the organisation and its employees.

- Observation only gives a snapshot view of an organisation.

3.3.6 Measuring

A variant of the observation technique is that of measuring. The distinction between the two techniques is made to highlight the primary use of observation as a technique to identify patterns and trends such as workload balance, whilst measuring implies a numerical activity in which specific values are assigned to the characteristics of an activity.

Examples of measures that may be taken are volumes, such as the number of staff within an organisation or the number of records held about a particular object or activity; rates, such as the number of orders processed per day and intervals, such the time between the arrival of a customer order and the despatch of the goods requested in that order. In each case, appropriate measures such as mean, maximum, minimum, mode and median values should be taken.

3.4 Preliminary Fact Documentation

In the succeeding chapters, we shall examine some of the contemporary techniques for modelling information system requirements, specifications and designs. However, before turning to some of these for formalised representations, it is worthwhile briefly examining the more informal fact documentation techniques which an analyst might find useful during the initial stages of analysing a system, in addition to notes, organisational structure charts (figure 3.2) and business system models (figure 3.4).

3.4.1 Matrices

Matrices or grid charts are useful documentation tools for recording the relationship between information, processes, departments and people. An example of staff duties in a hotel is shown in figure 3.8. Relationships between staff members and their duties are indicated by a cross in the table. Such a table might be useful to analyst for recording the various staff functions within an organisation for interviewing purposes.

STAFF \ DUTY	Front-desk	Reservations	Switchboard	Accounts office
S.James	x	x		
J.Hubbard			x	
F.Wilson	x	x		x
A.Scott	x			

Figure 3.8: A Staff-Duty Matrix

Another example of the matrix documentation technique is to record the origin and use of documents within an organisation, such as that shown in figure 3.9. In this case, the entries in the table vary to indicate the type of processing performed on the documents shown: C=create a document, R=read a document, U=update a document, D=delete a document.

PROCESS \ DOCUMENT	Customer account	Customer invoice	Sales bill	Receipt
Check-in	C			
Check-out	D	C	R	C
Sale	U		C	

Figure 3.9: A Document-Processing Matrix

3.4.2 Flowcharts

A variety of flowcharting conventions exist, each of which aim to document the sequencing of materials or information processing, decision-making or operational procedures within an organisation.

Figure 3.10 summarises the standard flowcharting symbols used by the *NCC (National Computing Centre)*, although a number of variations exist.

The flowcharting symbols can be combined to document clerical procedures or operations, such as the example of hotel customer reservation processing shown in figure 3.11.

The example shows the receipt of customer bookings, which are classified into those which can be serviced and those for which a room is unavailable. In the former case, the bookings are acknowledged and forwarded to a reservations clerk for further processing and the flowchart continues on page 2. Bookings which cannot be serviced are returned.

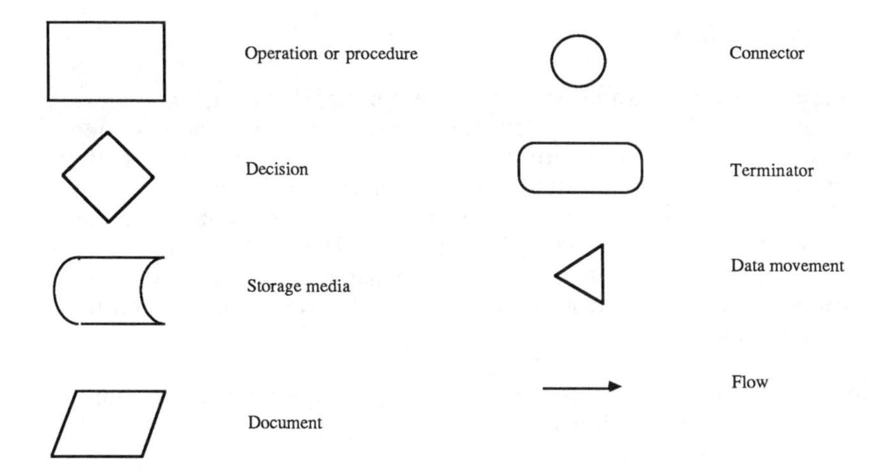

Figure 3.10: The NCC Flowcharting Symbols

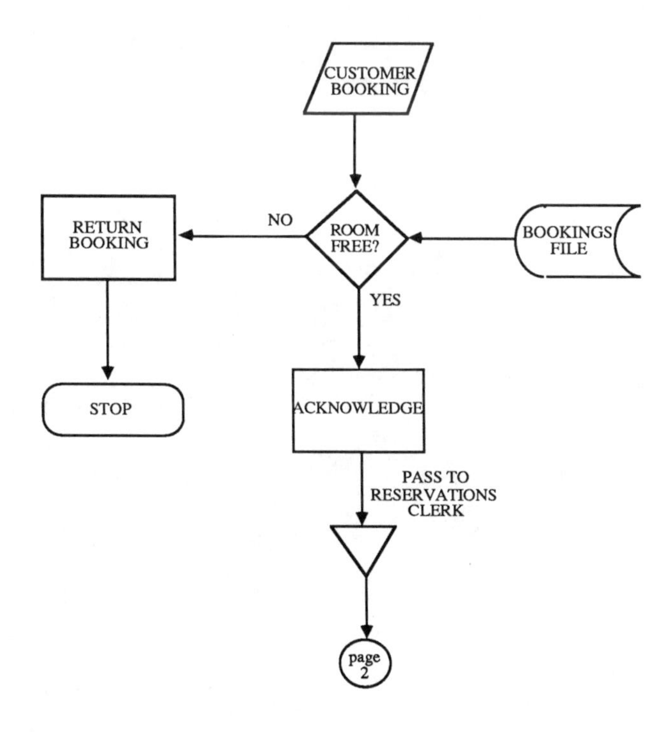

Figure 3.11: Flowchart for Customer Reservation Processing

3.5 Summary

The essence of any good system development is a full and correct understanding of the desired system and its requirements. The first step to this understanding is by determining the true nature and bounds of the problem faced by an organisation. Few formalised methods exist by which this can be achieved, but a variety of techniques exist including reading, interviewing, questioning and observation. These techniques can be combined to permit the analyst to build a comprehensive picture of an organisation, the environment in which it operates and problems and needs of the organisation.

Further reading on the subject of interviewing and questionnaire design can be found in *Moser and Kalton* (*1971*).

Chapter 4

Process Analysis

In the previous chapter, the main steps in problem determination and fact collection were outlined. However, in practice, neither of these tasks occur in isolation and indeed, fact collection is a continuous process, lasting at least through to design. In reality, fact collection and even problem determination, is conducted in a much broader framework of the systematic analysis of a system. The aim of this analysis is to organise the opinions, rules, procedures, policies and measurements gathered by an analyst into a coherent and ordered model of the required system.

This framework concentrates on two distinct aspects of a system: data and its processing. In this chapter we shall examine the concepts and techniques of structuring and representing information about processes within a system. To demonstrate these concepts and techniques we make use of a particular notation but, as noted in chapter 2, this notation is only an externalisation of principles and semantics shared by all process-oriented methods.

4.1 Introduction to Process Analysis

The objective of process modelling is to produce a specification that accurately embodies the requirements of future users of the system under development; these requirements having been distilled from the mass of fact and detail that the analyst obtains through interviewing, questioning, reading and measuring.

The desirable characteristics of any specification are as follows.

Coherence of Specification. The primary objective of the process modelling phase of the system development lifecycle is to produce a clear and understandable specification of what happens within a system. Thus the specification must be structured and presented in an understandable manner. This implies a systematic method of production and presentation, together with concise and commonly understood notation. This latter requirement is essential if full, user validation of a specification is to be performed. Furthermore, as a specification, the output should describe what is required of a system, rather than how it is to be provided.

Functionally Decomposed. A specification will inevitably need to represent a large number of different views of a system, in an integrated fashion. On the one hand, it must present a general overview, concentrating on the general functional areas of the system; whilst on the other hand, it must also describe the operational detail of the system. Thus in the case of a hotel administration system, the specification must both outline the general functions of catering, room booking and invoicing, as well as the detailed policy for room allocation and charging.

Graphical. Because specifications are typically large, it is essential that a specification is largely graphical. This will enable easier understanding by future users and analysts alike, compared to large volumes of written text.

Maintainable. The specification must be maintainable and thus should not contain duplication or redundancy.

The challenge to the analyst is to move from a problem definition to a single specification exhibiting the required characteristics. In principle, this is achieved as a two-stage process: initially by structuring a series of individual users' views of a system and subsequently through their integration into a single, system view. The remainder of this chapter is therefore concentrates on these two themes, beginning with user process modelling.

4.2 User Process Modelling

User process modelling is concerned with building a series of models describing how users see a system. In the simplest of systems, these step may almost be superfluous; but in larger systems it is a helpful stage in structuring user-supplied detail.

A variety of approaches exist to user process modelling , the most popular of which is a group of approaches known as *task analysis* (*Johnson et al, 1984*), whose purpose is to provide systematic ways to capture and record how individuals perform their tasks within a system, in terms of task steps, sequencing of tasks and conditions for repetition or selection of alternatives.

The techniques for representing this knowledge vary, but include traditional flowcharts and structure diagrams, such as those shown in figures 4.1 and 4.2 respectively. Both these figures show the detail of a task for accepting payment by credit cards. At the heart of the process is the need to obtain bank authorisation if the value of the transaction exceeds a certain limit.

In figure 4.1, each task step is shown as a rectangle and decision points shown as diamonds. The notation is a variation of that shown for flowcharting in chapter 3. In this particular manifestation, the chart is showing the flow of work, by indicating the tasks performed by an individual or process.

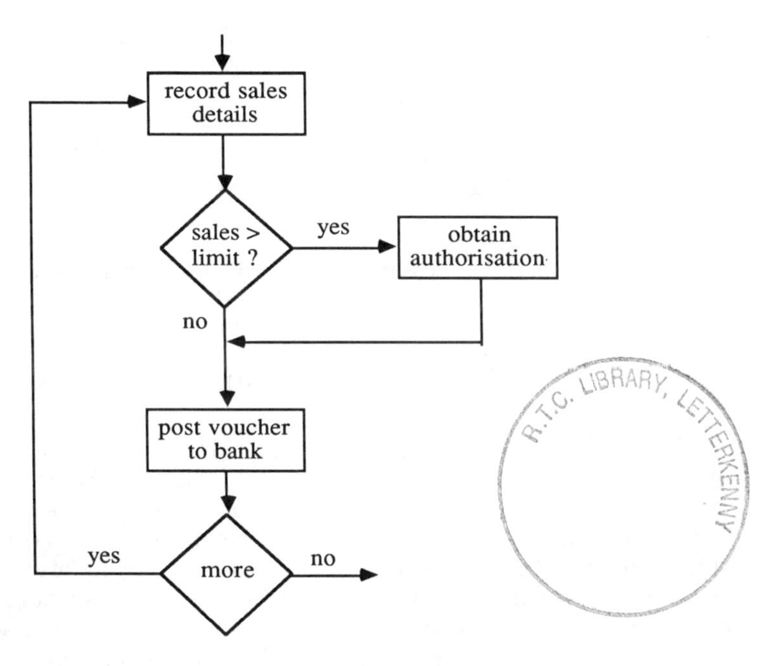

Figure 4.1: Authorisation task shown in a flowchart

In figure 4.2, the notation is that of a structure diagram. Each box represents a single or group task step, depending upon its position. If the box is a leaf on the tree structure it is a single task and if it is not a leaf, it represents a group of tasks. Boxes containing an asterisk, such *accept payment*, indicate repetition of the task group, whilst boxes containing circles in their top right-hand corner, such as *obtain authorisation*, represent a choice. The box containing a dash, represents a null alternative and thus indicates that the choice sequence is optional.

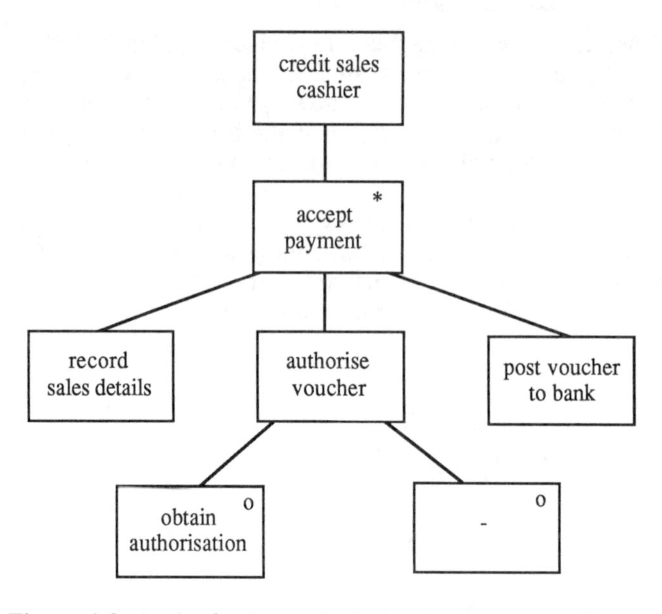

Figure 4.2: Authorisation task shown in a structure diagram

4.3 System Process Modelling

Having established individual user views of a system, the analyst must integrate these into a single, system view. In this book, we shall concentrate on the class of system development methods known as *structured methods*. Typically, commercial system development methods based upon the structured paradigm have numerous variations in notation or detailed application of the method. However, research has shown that there is a high degree of commonality between such methods (*Loucopoulos et al, 1987*) and therefore this book concentrates upon the general principles of the entire class of structured methods, rather than their detail.

The three main analysis techniques to be reviewed in this book are:

- Data Flow Diagrams, which show the interrelationships between processes in terms of the data that flows between them.
- Data dictionary definitions, which describe the elements of a data flow.
- Process specifications, which show the detail of processes.

Figure 4.3 shows the interrelationship between these components. At the heart of a system lies data. Typically, this data will have a well-defined structure which is described in a data dictionary. The data dictionary entries will describe the structure and characteristics of the data, together with commentary information obtained during the analysis phase of a project. Finally, the data held within a system is manipulated by a set of processes which fulfill the objectives of the information system.

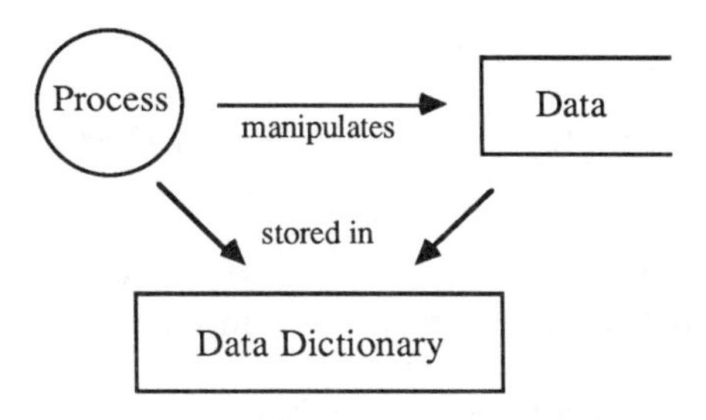

Figure 4.3: The Interrelationships Between Specification Components

4.3.1 Introduction to Data Flow Diagrams

Data flow diagramming is a technique which documents the inter-relationships between processes in terms of the data that flows between them, an example of which is shown in figure 4.4.

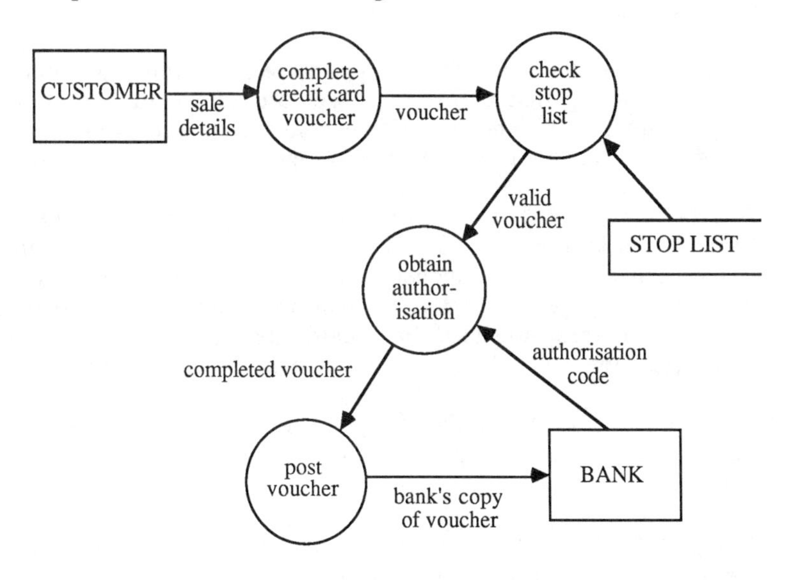

Figure 4.4: A Data Flow Diagram for Credit Card Authorisation

The diagrams are based upon the principles that the key components of a system are:

- a series of **processes**, each of which receive a flow of input data and transform that data into an output flow
- a series of **data flows** that pass into one process and out of another process
- a number of data repositories, known as **files**, which are available for access or amendment by a process
- a number of entities, outside the system, which pass data into the system (**sources**) or receive data from the system (**sinks**).

Through the combination of these four basic constructs, it is possible to construct a network representation of a system.

The specific data flow diagram terminology and symbols are as follows.

Processes. Processes are shown as circles in a data flow diagram and each process represents a detailed action that is performed within the overall system and which contributes to the aims and objectives of the system. A process transforms one or more inflows of data, from other processes, files or sources, into one or more outflows of data, to other processes, files or sinks.

Each process is labelled with a unique name and a reference number. It is worth noting that whilst processes are shown in this book as circles, a number of other notational conventions are used by other methods, including the use of ovals or octagons.

Data Flows. A data flow is a pipeline through which packets of information of known composition flow. Thus data flows represents an interface between other components of a data flow diagram. Most data will flow between processes, but it can also flow into or out of files and to and from sources and sinks. In all cases, the data flow is represented as a directed edge and is labelled with the name of the data flow.

It is important to stress that data flows must only represent data passing around the system and are not intended to represent the flow of control within the system.

Files. Files can be regarded as repositories of data and the term is used in its widest sense covering all forms of data storage such as tapes, discs, databases, filing cabinets, card indices etc.

Files are represented by an open-ended rectangle, containing the name of the file. Other variations included vertical or horizontal lines, with the name of the file alongside. A file will also have at least one data flow leading to or from itself. The direction of the arrow on these data flows is significant; an arrow pointing towards a file being used to indicate a net flow of data into the file, such as updating or addition, whilst an arrow pointing away from a file is used to indicate a net flow of data from the file, such as reading or deleting.

External Entities: Sources and Sinks. A source or sink is an entity which exists beyond the immediate scope of a system, but nevertheless is a net originator or receiver of data from the system. They are represented by squares, containing within them the name of the source or sink.

Thus the data flow diagram shown in figure 4.4 would be read as follows.

"Credit sale customers supply sales details which are entered onto credit card vouchers. When completed, the voucher is checked against a stop list . If the voucher is valid, an authorisation for the voucher is obtained. The Bank supply the appropriate authorisation code. Once the code has been obtained, the voucher is posted to the bank."

It is interesting to observe, even with this small examplc, how verbose written text can be compared to the precision and visual impact of the data flow diagram in figure 4.4 (allowing of course for the possible unfamiliarity of the reader with the data flow diagramming notation).

A moment's reflection on figure 4.4 will soon generate a number of questions about the system shown. For example, what happens when a voucher appears on the stop list or the bank refuse an authorisation code? These issues are returned to shortly.

4.3.2 Constructing Data Flow Diagrams

Data flow diagrams for small systems are normally easy to construct and often can be produced with little thought. However, for most large, commercial systems it is useful to adopt a simple set of guidelines which are designed to ensure the production of good data flow diagrams. These basic guidelines consist of the following.

- identify the static components within the system; i.e. those objects which contain data
- identify the main processes which consume and generate stored data

and the data flows between them

- expand and refine the data flow diagram
- review the diagram.

These guidelines are now examined in greater depth.

Identify the Static Components

During the initial steps of fact finding, the analyst will become increasingly aware of the major components of a system, such as the main data flows, the primary files that are maintained by the system and the chief, external influences on the behaviour of the system, i.e. sources and sinks. Thus the first step in the construction of a data flow diagram is to document the main structures in the system.

For example, in the hotel case study, the outline of a data flow diagram to show the overall processing of booking requests would probably look something like that shown in figure 4.5.

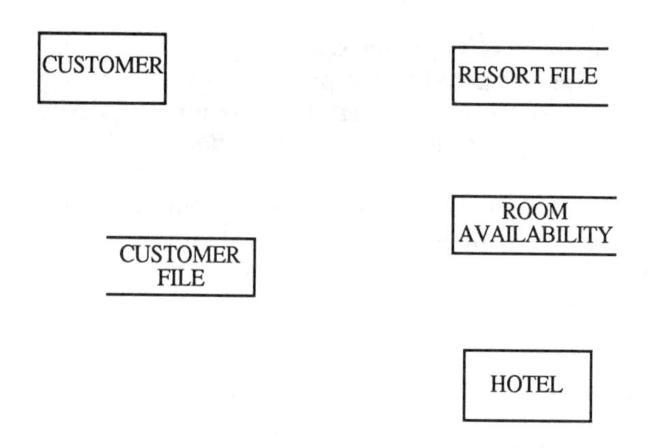

Figure 4.5: Outline Data Flow Diagram for Hotel Booking

The figure shows that there are two external entities, *customer* and *hotel*, together with three data files: *resort* file, containing details of hotels within a particular resort; *room availability* file, containing details of rooms within particular hotels and; *customer* file, containing details about customers, outstanding debts and discount arrangements.

Identify the Main Processes

Once the system boundary and main data files have been established, it is possible to start to add the main processes and data flows between them. This is achieved by identifying and drawing in the main pipelines or channels through which data passes, followed by additional of the main processes.

A good starting point for identifying data flows is to consider the main flows to and from sources and sinks. Once a data flow has been identified, it can be drawn onto the diagram and labelled, omitting at this stage the names of processes which receive or generate the flow.

When selecting data flow names, the following rules should be observed.

- Ensure that each data flow is named. If a name cannot be thought of, it is probably because the data flow either does not exist or consists of too many items of data in which case the flow should be split into two or more flows. Thus names such as *flow-1*, *input-flow* and *general-data*, would indicate a lack of understanding of the data and indicates the need for further analysis.

- Ensure that each data flow name refers to the whole data flow and not just part of it. For example, in figure 4.4, the external entity *customers* generate the data flow *sale details* which presumably includes all details about the customer, such as their name, address, credit card company and number, not just the value of goods sold.

- Words such as *form*, *list*, *information* and *data*, either alone or in conjunction with other names, should be avoided as they usually add nothing to the descriptive name of the data flow and probably show a misunderstanding of the true nature of the data flow.

Figure 4.6 shows the next stage in the construction of a data flow diagram for the hotel booking system and represents the basic data flows between the files and external entities, together with outlines of the main processing.

When identifying data flows, it is important to ensure that control information is not added to the diagram. The object of a data flow diagram, as the name implies, is to document the flow of data between processes within a system, not the manner by which these processes are controlled. Thus iteration, such as that shown in a task analysis description (figures 4.1 and 4.2), should not explicitly appear in an equivalent data flow diagram (figure 4.4).

Furthermore, when adding a data flow to a diagram, it is important to ensure that the flow is within the context or boundary of the system under study. Superfluous data flows that lie outside the system boundary, may appear to give extra information, but in reality add nothing to the proper understanding of the system.

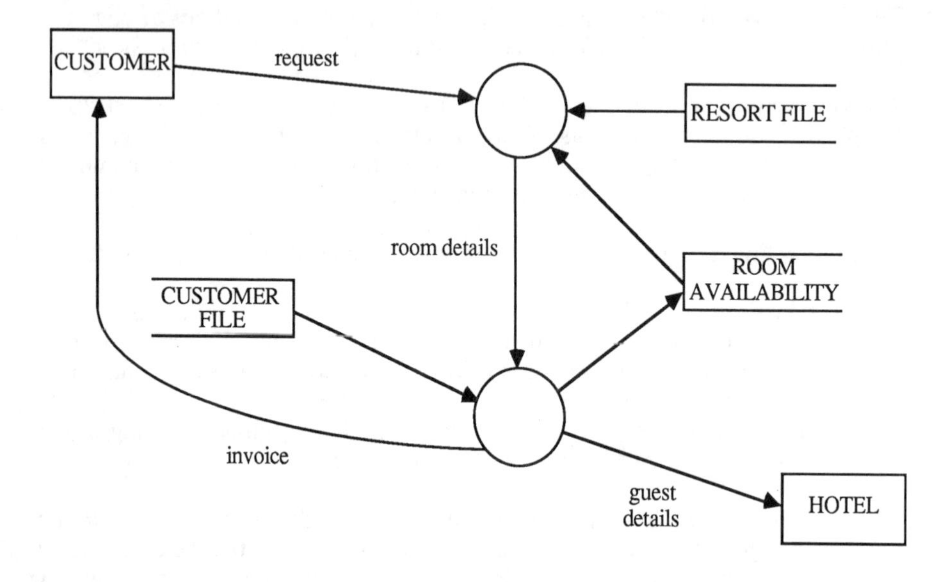

Figure 4.6: Hotel Booking System with Data Flows

For example, to add data flows to figure 4.6, concerning checking in and out of hotels would add nothing, if the system was purely about hotel bookings.

Expand and Refine the Diagram

Once the basic skeleton of a data flow diagram has been documented, it is possible to review each element of the diagram and through the use of follow-up interviews, questionnaires or reviewing existing material, expand and enhance the diagram. In the first instance, this will entail the naming of processes in the skeleton data flow diagram.

The easiest manner in which to proceed is to follow the paths leading from each source, naming each process, until arrival at a sink.

The general principle for naming a process is to base the name upon the input and output data flows. Thus, in figure 4.7 which shows the completed data flow diagram for the hotel booking example, the two process names shown explicitly reflect the fact that they are dealing with resorts and rooms.

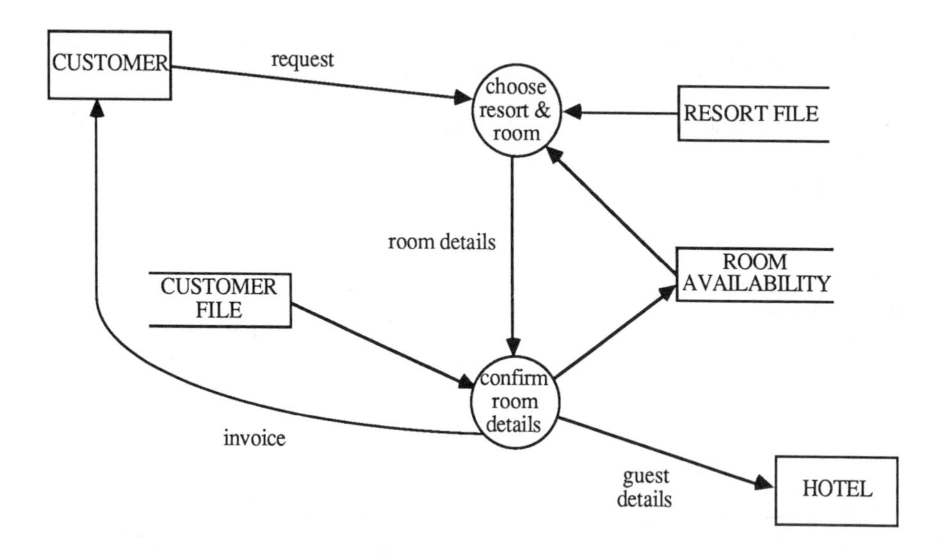

Figure 4.7: A Completed Data Flow Diagram for the Hotel Booking System

For example, the first main process is *choose resort and room* and not simply *choose.*

The rules which should be obeyed when naming processes are as follows.

- Names should attempt to cover all the activities that take place within a process. Names which cannot do this usually indicate that the process could be broken down into a smaller number of processes. Typically compound names, of the form *do-x-and-y* indicate this problem.

- Names should consist of a single strong verb and a singular object.

- Vague and imprecise names such as *process...* or *output...* should be avoided as they show that either the analyst understands insufficient about the process to devise a more precise name or that the process is too complex to be given a single name and thus should be split into smaller processes.

At the same time as naming the processes, the analyst should also be looking for opportunities to expand the data flow diagram. Clues as to the need for expansion are when the analyst has difficulties in assigning a name to a process. For example, the fact that an *and* is used in the first process of figure 4.7, is suggestive of the fact that the process could be split into two.

Another frequent area requiring expansion is usually wherever there is a process concerned with input from a user. It will almost be certain that such a process should be prefixed with another validation process.

Figure 4.8 shows an expanded data flow diagram for the hotel booking system. Notice in particular the additional request validation process and the expansion of the choosing and confirmation processes.

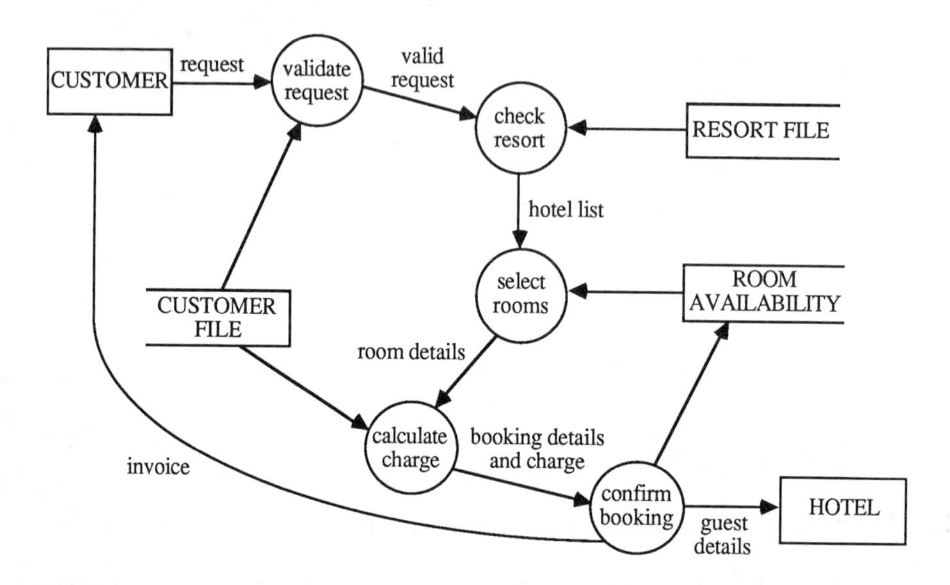

Figure 4.8: An Expanded view of the Hotel Booking System

At this stage it is important to still ensure that the diagram only contains major data flows that have a significant effect on the operation of the system. Thus there is no data flow that shows what happens to an invalid customer request.

Review the Diagram

Once a data flow diagram has been constructed, it is imperative that the diagram is reviewed. Analysis is an iterative process and it is not possible to accurately document a system first-time, because:

- the analyst has omitted detail which he has thought unimportant or has forgotten
- the user has omitted or forgotten detail

- a user has no knowledge of a particular activity and the analyst should have interviewed somebody else.

In the event of errors, the analyst may well find that it is easier to tear up any initial diagrams and start again.

4.3.3 Levelling Data Flow Diagrams

The strength of data flow diagrams lies in the fact that each diagram can literally be worth a thousand words. However, for all but the most trivial information system, more than one sheet of A4 paper will be required, immediately creating a management problem and devaluing the worth of the diagrams. What is required is a technique by which large systems can be split down into sensible sized, logical units, allowing quick reference to any part of the system description. The technique employed in data flow diagramming and which achieves the desired effect is known as levelling and represents a functional decomposition approach to system specification, in which a system is described in increasing levels of detail.

The Concept of Levelling

The concept of data flow diagram levelling can best be explained through two observations.

First, consider an analyst who has constructed a standard data flow diagram. Before the diagram can be regarded as complete, the analyst must validate the diagram. Ideally such validation would be carried out by those individuals who contributed to the analyst's knowledge of the system, that is those people whom the analyst interviewed, observed and questioned. However, if a large number of individuals were involved, ranging from senior management to shop-floor workers, validation is not easy because the view that a senior manager has of a set of activities is likely to be much broader and more general than the narrow, but detailed view, that a shop-floor worker will have. A senior manager may be confused by a mass of fine detail, as would a shop-floor workers confronted with a description of parts of the system about which they knew nothing. What is required is a set of integrated data flow diagrams capable of presenting a broad, high-level view of a system and a series of more restricted, detailed views of parts of a system.

The second observation concerns the use of a data flow diagram by a system designer, or indeed by analysts other than the author. Like the senior manager, when first confronted with a system specification, a designer does not want to be presented with a mass of detail about the system, but instead requires a complete and general overview; only later will the detail be required.

Levelling, therefore, is based upon the philosophy that underlies these two observations, namely that the sensible way in which to decompose a system is through having a series of diagrams. The first diagram is used to present an overview of a system, showing only its major components. Subsequent diagrams are then used to specify increasing levels of detail about different aspects of the system.

For example, consider figure 4.9 which shows a high-level view of the hotel booking system. Note that the diagram, of necessity, does not describe the problem area in sufficient detail. Indeed, it would be wrong if it did so.

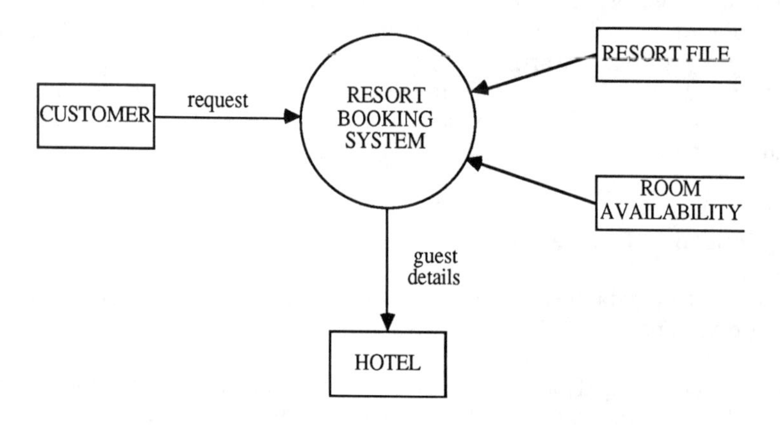

Figure 4.9: Overview of the Hotel Booking System

Nevertheless, the diagram provides the reader with an immediate feel as to the purpose of the system, its principal user and the recipient of the information generated by the system. The diagram in figure 4.7, by contrast gives a more detailed view of the system and the diagram in figure 4.8 gives the most detailed view. This concept of increasing detail is known as *levelling* and is an important concept when describing large systems.

In practice, analysts do not tend to show everything on a single diagram, since it would simply become bigger and more complex. Instead, it is the practice to divide a system up into a number of partitions which are documented in separate diagrams and refined into increasingly levels of detail. Thus rather than constructing the single, detailed diagram shown in figure 4.8, an analyst would construct a general overview of the system, such as those shown in figures 4.7 and 4.9, and supplement these with additional, partial diagrams detailing individual processes within the system. For example, the first process in figure 4.7, *choose resort and room*, might be further elaborated by the diagram shown in figure 4.10, in the same way that figure 4.6 is an elaboration of figure 4.9.

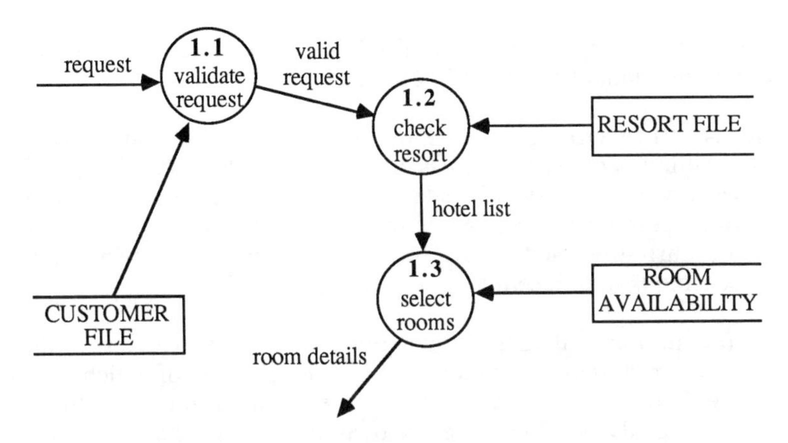

Figure 4.10: A Data Flow Diagram for Choosing a Resort and Room

This collection of data flow diagrams, is known as a levelled set of data flow diagrams and the general hierarchical relation between each diagram is shown in figure 4.11.

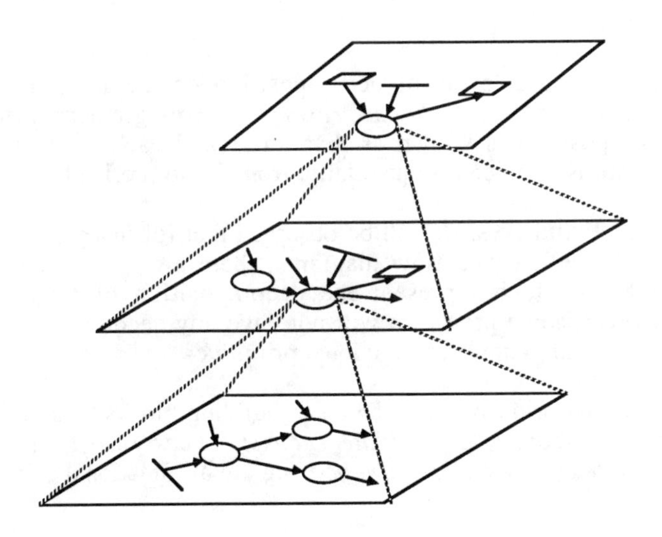

Figure 4.11: The Structure of a Set of Levelled Data Flow Diagrams

Figure 4.12 shows the full set of levelled data flow diagrams for the hotel booking system example.

Levelling Conventions

When constructing a set of levelled data flow diagrams, a number of conventions should be followed. These are as follows.

Process Referencing. One of the objectives of data flow diagram levelling is to permit the easier handling of large system specifications. However, as the sample data flow diagrams stand at present, reference to processes and diagrams is somewhat cumbersome, since only names are currently used. Therefore a hierarchical numbering and naming scheme is introduced, as shown in figure 4.12.

The high-level diagram which gives a complete overview of the system is referred to as the context diagram the purpose of which is to specify the domain of study and show the net inputs and outputs to the system under analysis. The process shown in the context diagram is known as process number 0, although this is often omitted from the diagram.

The first level of decomposition of the context diagram is referred to as diagram 0 since it is a decomposition of process 0. Each process within this diagram is numbered in the form *0.n* where n is an integer from 1 upwards. Again, because all diagram 0 processes commence with a 0, the leading 0 and point is usually omitted from the process number.

Where further levels of decomposition occur, as is the case with process number 1, *choose resort*, the diagram representing the decomposition is known as diagram 1 and each process within the diagram is numbered sequentially, from 1 and prefixed by *1.n*.

Functional Primitives. It will be observed that for many processes in the set of levelled data flow diagrams, there are subordinate data flow diagrams which represent the decomposition of those processes. However, some processes will not have any decomposition and these processes are known as functional primitives.

Functional primitives are the basic building blocks of a entire system. Thus in figure 4.12, *validate request, check resort, select rooms, calculate charge* and *confirm booking* are all functional primitives.

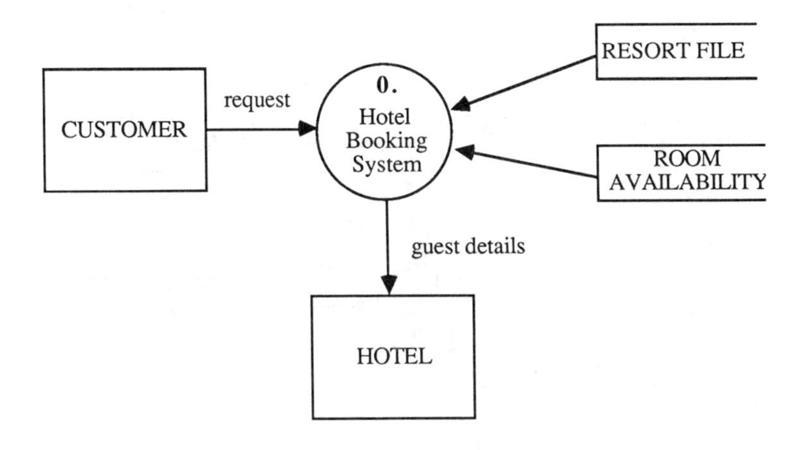

Hotel Booking System: Context Diagram

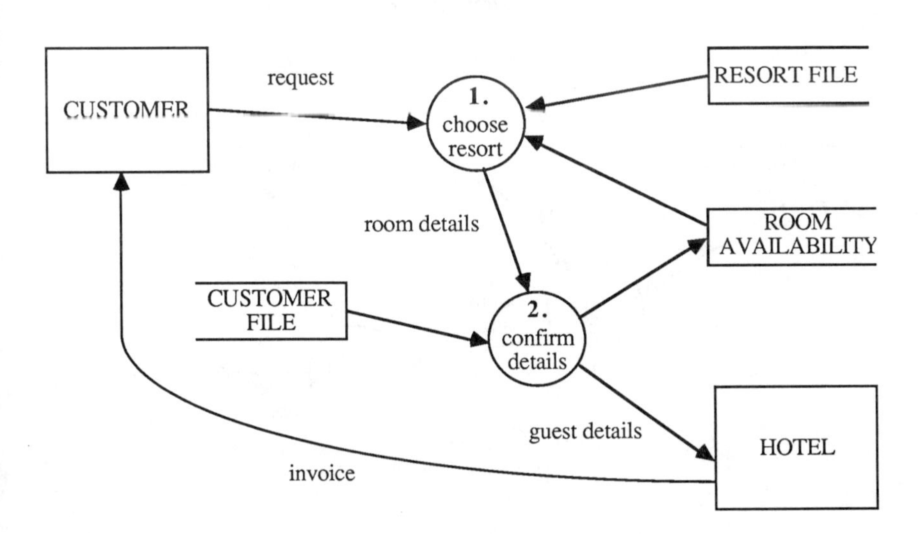

Hotel Booking System: Diagram 0

Figure 4.12(a): The Hotel Booking System

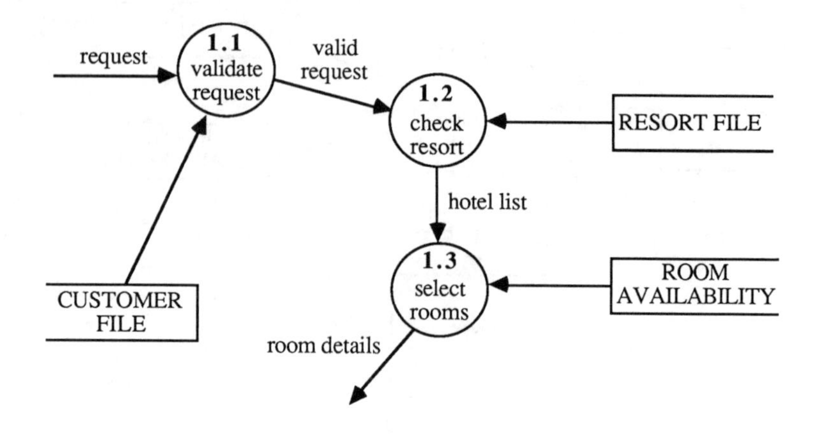

Hotel Booking System: Diagram 1

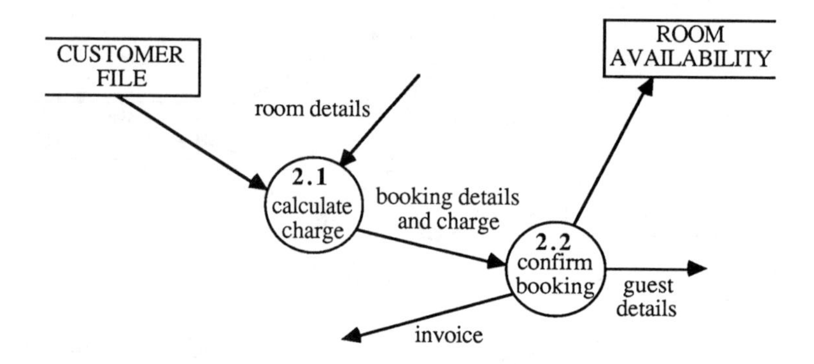

Hotel Booking System: Diagram 2

Figure 4.12(b): The Hotel Booking System

Because functional primitives are those processes which perform work within a system and because data flow diagrams provide no further mechanism by which this work can be specified, each functional primitive has associated with it a *process specification* and these are described in detail in the following section.

Balancing. The rule of balancing states that the data flows into and out of a given process must also appear on any data flow diagram that is a decomposition of that process. For example, in figure 4.13(a), the levelled set of data flow diagrams are balanced because the data flow X into process 2 is also shown in diagram 2 and the data flows Y and Z which flow out of process 2 are also shown in diagram 2. However, in figure 4.13(b), the set is not balanced since there is an additional out flow in diagram 2 which is not shown on the higher level diagram and the flow Y is missing from the lower level diagram.

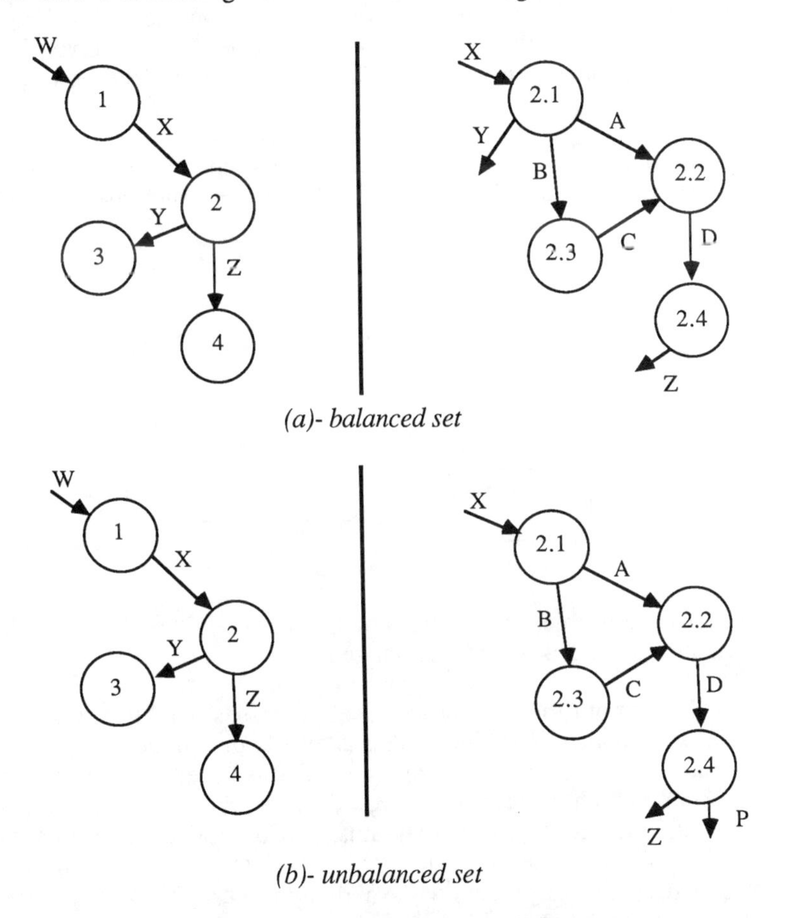

Figure 4.13: Levelled Data Flow Diagrams

4.3.4 Hints on Data Flow Diagramming

Annotation of Diagrams

Some textbooks and practitioners often advocate the additional annotation of data flow diagrams. For example, one technique is to place the symbols * or + between data flows, the former symbol indicating that the process requires both data flows before it can take place, whilst the latter symbol is used to indicate that only one of the two data flows is required. This practice should be avoided since it tends to force the analyst to begin to think about a system at a procedural level and to lose sight of his task to document data flows.

Another technique that should be avoided is the labelling of the source and destination of data flows in lower level data flow diagrams. For example, in figure 4.13(a), it would be wrong to label the flow Z on diagram 2 as flowing into process 4, since this has already been documented on diagram 0 in the set and should the process number 4 ever be renumbered would require two amendments to the diagrams instead of one- leading to possible inconsistency.

Another problem with marking diagrams occurs where special symbols are used to indicate duplicated external entities or files, such as the following symbols:

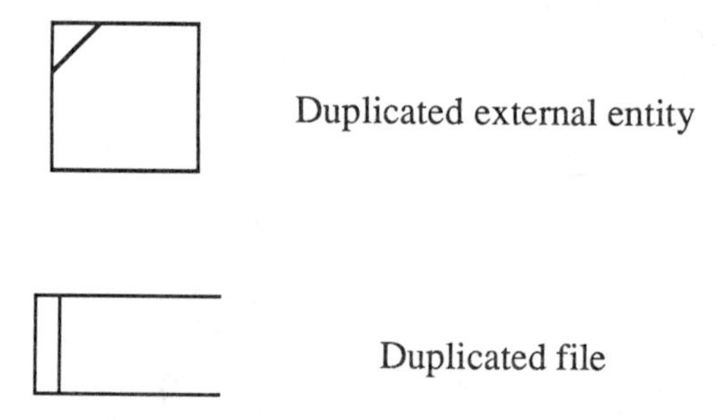

Again there is a problem of diagram maintenance and consistency when an external entity or file is added or removed.

Finally, the temptation to mark those processes which are functional primitives should also be avoided. This is mainly due to the fact that at some future time they too may be decomposed and thus necessitate the removal of the marking, which if forgotten, would cause the diagrams to become inconsistent. Indeed, data flow diagrams should be levelled so that when a process is sufficiently simple, the details of that process can be entirely encapsulated within the process name and any reader should have no wish to look for a further decomposition.

Extent of Decomposition

It is often a question of personal judgement how far a system should be decomposed. As a general guideline, decomposition should continue until the specification of a process can be written on one side of a piece of A4 paper, thus keeping the specification to manageable proportions.

Extent of Partitioning

Decomposition refers to the depth to which we describe a system. Partitioning on the other hand refers to the breadth by which a system is divided. Notice that in the levelling of the hotel booking system outlined in figure 4.12, the first level of decomposition of the context diagram, diagram 0, shows the entire system. However, at the second level of decomposition, the decomposition of processes 1 and 2 are shown on different diagrams, rather than on a single diagram. This is because with a single diagram, too much would have to be fitted into one diagram. A good rule of thumb is to aim to get between 5 and 9 processes on any one diagram.

Checks for Usefulness

A check should be made on all data flow diagrams to ensure their usefulness. Complex or irrelevant data flow diagrams probably means that the analyst has failed to understand the system and thus require further work. The following are tell-tail signs:

- process names such as *handle input* or *generate output* indicate a lack of understanding of a process

- data flows such as *input items*, *various data* or *command code* (which is control information) shows a misunderstanding of a process

- crossing data flows indicates the need for the division on the diagram into a series of separate diagrams (known as partitioning)

- complex interfaces in which functional primitives have a large number of data flows in and out indicate the need for further decomposition

- uneven partitions in which some diagrams have only one or two processes, whilst others have very large numbers of processes suggest potential partitioning problems.

4.4 Data Dictionaries

4.4.1 Purpose

A data dictionary is a repository of data about data. In terms of analysis, data is kept principally about the following three types of item.

Data Flow and Data Item Specifications. Each data flow appearing on a data flow diagram must have an entry in the data dictionary. This entry is used to describe the decomposition of data flows either in terms of other data flows, which appear in lower level data flow diagrams, or in terms of data items, which are individual packets of data generated and consumed by processes at either end of the data flow. Each data flow and data item specification is expressed in a data specification language, examples and the structure of which is described below.

Data flow and data item specifications are often supplemented by information relating to the characteristics of the data such as volume, frequency and security and thus assisting the system designer in his task of selecting suitable computer media, although these characteristics are usually not apparent until towards the end of an analysis.

File Specifications. The data dictionary is also used to hold information about the structure of files and the data which appears within them and are expressed in a data specification language. Where a system has more than one file, it is also necessary to analyse the relationships between files and to construct a data structure diagram. This is described fully in the chapter on data analysis.

Process Specifications. Process specifications document how incoming data flows are transformed into outgoing data flows. Every data flow, data item and file which the specification references must have a corresponding entry in the data dictionary. The process specifications may be documented in any form of notation, although semi-formal techniques such as decision tables or Structured English are preferable to standard, written English because of their relative conciseness and lack of ambiguity. These techniques will be discussed later.

Thus the contents of the data dictionary, together with a levelled set of data flow diagrams can be used to specify a complete system. Before looking at the data dictionary entries in more detail and the techniques used to represent the information, it is first worth identifying the characteristics that any representation form must have. These are as follows.

- Data dictionary entries must be precise and concise, data flow, data item, file and process specifications must pinpoint the exact characteristics of each type of item, leaving no room for ambiguity and it is for this reason that natural language is ruled out as a contender for the language to be used in a data dictionary

- Data dictionary entries must not contain redundancy since this leads to an increased effort when applying updates to the dictionary and, if forgotten, will cause the dictionary to become inconsistent.

4.4.2 Data Specification Language

Each data flow in a set of diagrams must be analysed and specified in the data dictionary in a top-down fashion. In practical terms this means that each data flow is represented in terms of other data flows or data items; where a data item is regarded as an atomic unit whose definition is totally encompassed by its name and for which no further decomposition is meaningful.

Because a data flow specification may contain subordinate flows or data items which are mutually exclusive, repeated or optional, the following notational conventions are used:

IS EQUIVALENT TO	=
AND	+
EITHER..OR	[ ]
ITERATIONS OF	{ }
OPTIONAL	( )

For example, a data item called room details that consisted of a hotel name and a series of room numbers, would be represented as:

room details = hotel name + {room number}

An example of a selection (either...or) would be:

amount due = [dollar amount, sterling amount]

In general all names referenced on the right-hand side of a specification, also have a specification elsewhere in the data dictionary. However, at some stage, the definitions of items such as age, name or address becomes impractical and unnecessary, since their specifications are obvious. Thus at this point, the chain of specifications may stop.

Where a data flow or data item is known by an alias, it is unlikely that an analyst can ever eliminate their use within an organisation and so should be documented by the addition of further specifications, such as:

account number= bank ID + customer number + check digit

= customer banking number

in which account number is composed of bank ID, customer number and a check digit and has the alias customer banking number.

Whilst aliases represent redundancy, they will however eventually be removed from the data dictionary once standard terms for a future system have been established.

4.5 Process Specifications

Processing and control information which has been omitted from a data flow diagram belongs in a process specification. Each functional primitive, that is, each process which has no decomposition, has one process specification within the data dictionary. These specifications can be represented in a variety of languages, Structured English and Decision Tables being the most popular.

4.5.1 Structured English

Structured English is a rigid subset of the English language which omits adjectives and adverbs, compound and complex sentences, all verb modes except imperative and most punctuation. The result is a language containing a limited set of conditional and logic statements with nouns and strong verbs.

Like many other specification languages such as pseudo-code there is no inclusive definition and standards vary between organisations, but nevertheless, the objectives of conciseness, preciseness and the lack of ambiguity apply to all variants. However, Structured English possesses the three standard control constructs of sequence, selection and iteration which together with primitive actions permit the specification of any system.

Primitive Actions

Primitive actions are used to inform the reader of something which must be done, as opposed to when it is to be done. Primitive actions are expressed in Structured English as imperative statements and should have the following characteristics:

- statements should be concise avoiding vague words such as process or handle
- statements should contain a strong verb that clearly defines the required function
- the object of the statement should be stated explicitly and selected from the data dictionary, e.g.:

READ-FILE STOCK-DETAILS

where READ-FILE is the verb and STOCK-DETAILS is a file name appearing in a data dictionary.

Control Constructs

Structured English recognises the following three control constructs.

Sequences. Sequences represent one or more actions which take place, in sequence and without interruption. In Structured English, a sequence is defined by the successive appearance of a set of primitive actions.

Selections. A selection is a construct in which there exists a series of alternative policies from which exactly one policy is selected. Selections are represented in the form:

```
IF <condition>
   <statements>
ELSE
   <statements>
```

where IF and ELSE are special Structured English keywords, <condition> is a relational condition and <statements> are further Structured English control constructs or primitive actions.

In the case where more than one alternative may require selection, a case-type construct can be used of the form:

```
CASE
   WHEN <condition>  <statements>
   WHEN <condition>  <statements>
     :       :             :
```

Iterations. An iteration is where a policy or series of actions is repeated within some bound. Iterations can be represented either by a

DO...WHILE construct or a REPEAT...UNTIL construct.

In the case of a DO...WHILE construct, which has the following structure:

```
DO WHILE <condition>
    <statements>
```

the condition is first checked and, if true, the statement body must be performed. By contrast, the REPEAT...UNTIL construct, of the form:

```
REPEAT
    <statements>
UNTIL <condition>
```

ensures that the statement body is executed at least once, before the condition is checked.

The strength of these constructs lies in the fact that they all have a single starting point and a single finishing point, resulting in a structured and readable document when combined.

The following example shows the use of all control constructs.

```
total_charge = 0
REPEAT
    get_next_room
    IF    room_type = "EXECUTIVE"
       total_charge = total_charge + £57
    ELSE
       total_charge = total_charge + £34
UNTIL all_booked_rooms_processed
          OR total_charge > credit_limit
```

The advantages of Structured English are:

- it is concise and precise, allowing easy reading and avoiding ambiguity and misunderstanding
- the language notation used may be tailored to suit the user, providing of course that it does not infringe the general principles of Structured English
- there must exist a cross-referencing with any data flow diagrams and data dictionary entries, thus permitting thorough verification.

Unfortunately, Structured English also has its disadvantages, but these can be overcome. The main disadvantage is that the apparent formality of Structured English is somewhat alien when first read or written. It is very likely that a novice will struggle with it at first, but the effort is worthwhile. As far as formality goes, one must avoid the pitfall of assuming anything written in Structured English is correct. The medium is not formal and cannot prove correctness, it is simply terse and to the point!

4.5.2 Decision Tables

A decision table is a tabular representation of conditions and actions and an indication under which conditions, which actions must be performed.

A decision table consists of four quadrants, as shown in figure 4.14.

- the condition stub contains a list of all possible conditions that can arise within the process
- the action stub contains a list of all the possible actions that can occur within the process, appearing in the desired order of execution
- the rules quadrant contains selectors which identify different combinations of the possible conditions
- the action entries quadrant contains indicators which select those actions to be performed.

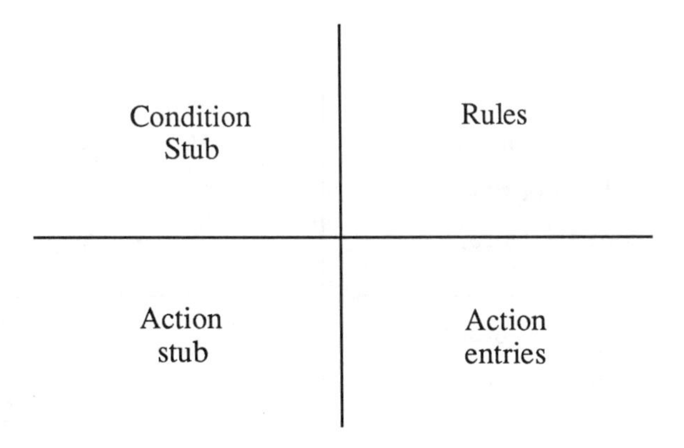

Figure 4.14: Structure of a Decision Table

Given this basic structure of decision tables, three variants are possible:

- a limited entry decision table
- an extended entry decision table
- a mixed entry decision table.

Limited Entry Decision Tables. A limited entry decision table contains only the binary selectors Y and N and the catch-all selector - in the rules quadrant and the action selector symbol X in the action entries quadrant.

Figure 4.15 shows an example of a limited entry decision table. This table, which documents how an order is validated, indicates that three conditions are possible, namely that the customer has satisfactory credit, that the customer is a prompt payer and that special clearance can be given to an order. The table also shows that on the basis of different combinations of conditions, an order can be accepted or rejected.

The rules and action entries quadrants are used to identify under what conditions what actions must take place. Thus rule 1, which is identified as the first column in the rules quadrant, shows that provided that a customer has satisfactory credit (indicated by Y), irrespective of whether they are a good payer or have received special clearance (indicated by -), the order is to be accepted.

	1	2	3	4
Credit satisfactory	Y	N	N	N
Prompt payer	-	Y	N	N
Special clearance	-	-	Y	N
Accept order	X	X	X	
Return order				X

Figure 4.15: A Limited Entry Decision Table

Rule 2 states that in the case where a customer does not have satisfactory credit (indicated by N), but nevertheless is a prompt payer, then irrespective of whether they have obtained special clearance, the order is to be accepted. And so on until all eventualities are covered.

Mixed Entry Decision Tables. In a mixed entry decision table, such as that shown in figure 4.16, is a table in which indicators other than an X appears in the action entries quadrant. In the table shown in figure 4.16, the action pay is qualified by one of the indicators in the action entries quadrant. Thus in rule 1 where an employee is not salaried and has worked more than 40 hours, pay is qualified by *overtime rate*, indicating that pay is on an overtime basis, as opposed to the alternative cases of *regular rate* or *regular salary*.

	1	2	3
Salaried employee	N	N	Y
Hours worked > 40	Y	N	-
Pay	Overtime rate	Regular rate	Regular rate

Figure 4.16: A Mixed Entry Decision Table

Extended Entry Decision Tables. In an extended entry decision table, selectors in the rules quadrant are no longer simply binary (Y or N), but may take on specific values or ranges of values.

An example of an extended entry decision table is given in figure 4.17. As in limited entry decision tables, the condition *approved credit* is selected using the binary selectors true (Y) or false (N). However, rather than document three order quantity conditions (0-25, 26-55 and greater than 56) in the condition stub, a single condition, *quantity ordered* is used and the rule selector takes on value ranges to identify specific order quantities. This technique permits more a compact representation of a decision table in which there are a large number of value-related conditions.

	1	2	3	4
Approved credit	N	Y	Y	Y
Quantity ordered	-	0-24	25-55	56-99
Discount (%)		0	5	10
Release order		X	X	X
Reject order	X			

Figure 4.17: An Extended Entry Decision Table

Validating Decision Tables

One of the advantages of limited decision tables is that a completeness check can be applied to ensure the correctness of the table. The check can be applied using the algorithm:

1. Defining c to be the number of conditions in the condition stub and calculate: 2^c

2. For each rule, scan down the rule counting the number of catch-all indicators. Let the number of catch-all indicators be n. Calculate the value b as: 2^n

3. Add all the bs together and this value should be the same as the value obtained in step 1. If the values are equivalent, the decision table documents all cases. If the values are different, then there are either some rules are duplicated or some rules are missing.

Advantages of Decision Tables

The main advantage of decision tables is that there are easily encodable in industry standard languages, such as COBOL. For example, the extended entry decision table in figure 4.17 can be directly translated into the COBOL EVALUATE statement shown in figure 4.18. As well as such manual translation, decision table translators exist for automatic translation.

Other advantages of decision tables include the following:

- decision tables are easily understood

- alternatives are shown side by side
- the cause and effect relationship is shown, thus permitting easier user validation
- it is possible to check that all combinations of conditions have been considered.

```
EVALUATE  approved-credit  ALSO  quantity-ordered
   WHEN          N         ALSO     ANY
           PERFORM reject-order
   WHEN          Y         ALSO     0 THRU 24
           MOVE 0 TO discount-rate
           PERFORM release-order
   WHEN          Y         ALSO     25 THRU 55
           MOVE 5 TO discount-rate
           PERFORM release-order
   WHEN          Y         ALSO     56 THRU 99
           MOVE 10 TO discount-rate
           PERFORM release-order
END-EVALUATE
```

Figure 4.18: A Decision-Table Encoded in COBOL

4.5.3 Decision Trees

An alternative to decision tables are decision trees and these are mentioned because of their graphical properties.

Decision trees employ tree structures which show the conditions and actions within a problem. The advantage of decision trees is that they are very easy to understand and there is no need for special training.

Figure 4.19 shows an example of a decision tree. At the root of the tree appears the name of the process *Account Policy*. Each branch of the tree is labelled with a condition and the leaves of the tree are labelled by actions which must be performed if all the conditions between the leaf and the tree root are true.

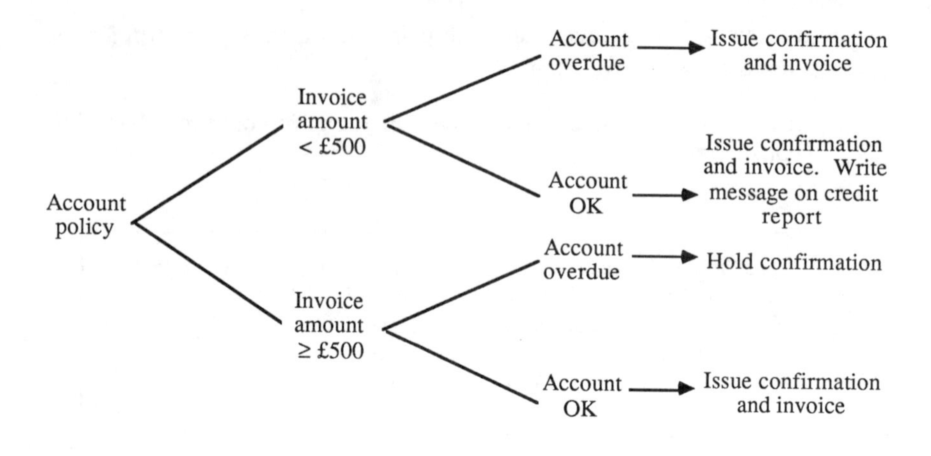

Figure 4.19: A Decision Tree

4.6 Summary

This chapter has shown one of the popular techniques for system specification. A system is described using three modes of representation:

- data flow diagrams, which show the relationships between processes, data flows, data repositories and external influences beyond the boundary of the system in a graphical and structured manner

- process specifications, using Structured English or decision tables, which document in a precise and concise fashion the elementary processing operations that are performed by each functional primitive in a set of data flow diagrams

- data dictionary entries, which describe the nature of data flows and hold the process specifications.

Further references to the techniques covered can be found in *de Marco (1978)*.

In the next chapter, attention is focused on the modelling of the data used by the system and the relationships between the data.

Chapter 5

Conceptual Data Modelling

In chapter 4, development of an information system was considered from the point of view of a system's behavioural aspects, that is the *processing* part. As already discussed, an information system must be examined in terms of both the processes which operate within the system and in terms of the underlying data.

Data is a major component of any organisation and therefore the modelling of corporate data, and its relationship to processes, is of paramount importance in the development of effective information systems. This chapter is concerned with the activity of understanding, documenting and analysing corporate data for the purpose of developing information systems. This activity is known as *conceptual data modelling* or simply *data analysis*.

Data analysis is concerned with taking an amorphous mass of facts about the data used within a system and turning them into a precise, unambiguous and non-redundant data description which would then serve as the basis for a database implementation.

5.1 The Conceptual Schema

The recognition of data as an important organisational resource and the subsequent developments in database technology have been the main factors behind the introduction of *conceptual data modelling* techniques as a viable alternative to ad-hoc data design. These techniques encourage the analyst to think in terms of the structure and meaning of the data in a way which is machine-independent, rather than in terms of specific file or database structures. Within the context of information systems development the activity of conceptual data modelling and its related activities are shown in figure 5.1.

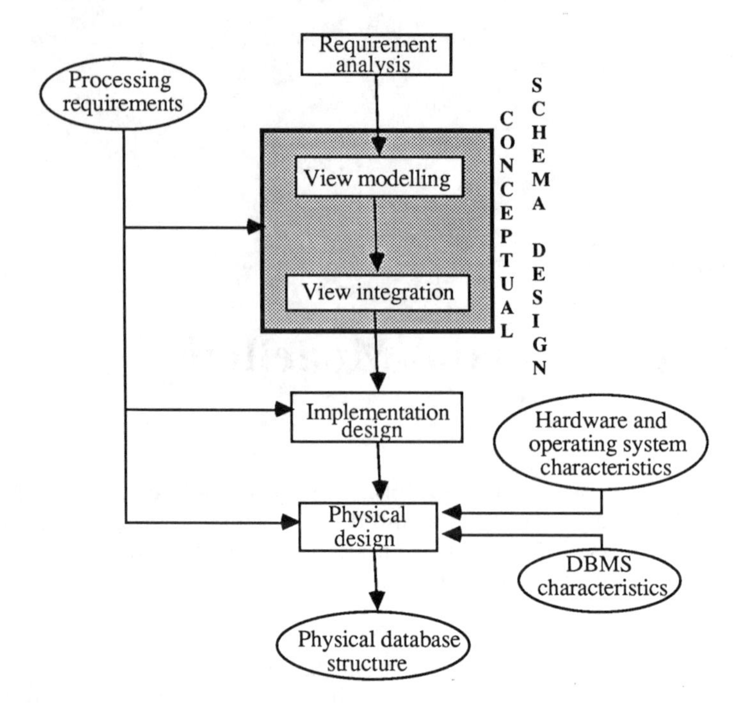

Figure 5.1: Conceptual Data Modelling

The objective of conceptual data modelling is to determine the data requirements of an enterprise, specify these requirements so that all processing requirements may be satisfied and eventually build a database structure which implements the specification of the data model.

The importance of data analysis in the development process has been recognised by many researchers and practitioners. The work of *Codd* (*1970*; *1972*) and *Chen* (*1976*; *1977*) and most importantly the recommendations of the ANSI/SPARC committee for the design of databases (*ANSI, 1975*), have been key milestones in the field of data modelling.

Conceptual data modelling emphasises the fact that an organisation deals with many different types of data which may have complex interrelationships. Before any attempt is made in designing and creating computer files, a thorough understanding of the data and its relationships is required.

Following the recommendations of ISO (*van Griethuysen et al*, *1982*), a domain specification, derived from the use of a conceptual modelling formalism, is known as a *conceptual schema* and represents abstractions, assumptions and constraints about that domain. More accurately it represents a user's conceptualisation of the application environment.

The primary use of a conceptual schema is in understanding a specific application domain. Naturally, this activity involves communication with users of the modelled application and, since the communication is to be carried through the conceptual schema, the schema must be *cognitive* in nature. In other words the schema's concepts must be relevant to the milleu in which the information system is used, (i.e. the object system), and not related to its design or implementation. Moreover, the schema should force active participation of users by stimulating and generating questions as to how reality is abstracted and assumptions are made.

Therefore, a conceptual schema is used to harmonise and facilitate human communication. It may be regarded as a general agreement between all persons concerned in the development of an information system and in particular between developers and end-users. This agreement corresponds to the way in which the universe of discourse is perceived at some point in time. In addition, a conceptual schema should be capable of supporting *all* applications in the universe of discourse over their lifetime. In other words a conceptual schema should determine the kind of information which will be found in a database which is integrated and shared by many different applications. This idea is shown in figure 5.2, in which a single, global view of the data is shown being shared by two different applications.

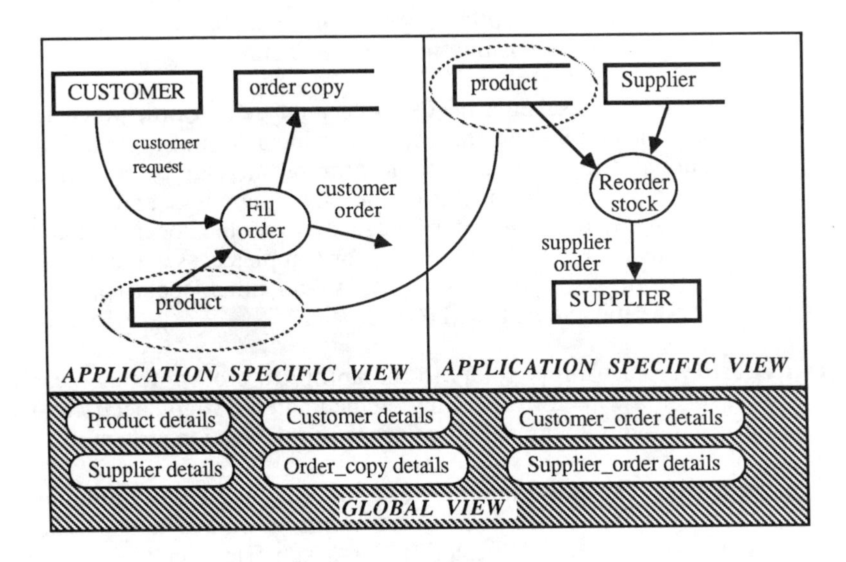

Figure 5.2: A Conceptual Schema is a Global View of Data

A number of desirable properties for a conceptual schema have been proposed (*Balzer & Goldman, 1979*; *Yeh, 1982*; *van Griethuysen et al, 1982*; *Nijssen & Duke, 1987*; *Olle, 1988*; *Kerola, 1988*) and can be summarised as follows.

Implementation Independence. No implementation aspects such as data representation, physical data organisation and access, as well as aspects of particular external user representation, (such as message formats, data structures, etc) should be included in a conceptual schema.

Abstraction. Only general aspects of an information system and the universe of discourse should be represented (i.e. those not subject to frequent change). Abstraction results in a schema in which certain details are deliberately omitted.

Formality. Formality implies that descriptions should be stated in an unambiguous syntax which can be understood and analysed by a suitable processor. The formalism should be based upon a rich semantic theory that allows a clear relationship between descriptions in the formalism and the world being modelled (*Mylopoulos, 1986*).

Constructability. A conceptual schema should be constructed in such a way as to enable easy communication between analysts and users and should accommodate the handling of large sets of facts. In addition, a conceptual schema needs to overcome the problem of complexity in the problem domain, by following appropriate abstraction mechanisms which permit decomposition in a natural manner.

Ease of Analysis. A schema needs to be analysed in order to determine whether it is ambiguous, incomplete, or inconsistent. A specification is ambiguous if more than one interpretation can be attached to a particular part of the specification. Completeness and consistency require the existence of criteria against which the specification can be tested. However, the task of testing for completeness and consistency is extremely difficult, normally because no other specification exists against which it can be tested (*Olivé, 1983*).

Traceability. Traceability refers to the ability to cross-reference elements of a schema with corresponding elements in a design specification and ultimately with the implementation of an information system.

Executability. The importance of executability is in the validation of a schema (*Balzer et al, 1983*; *Karakostas & Loucopoulos, 1988*). In particular it refers to the ability of a specification to be simulated against relevant facts in the modelled reality. The executability of the descriptions in a schema is subject to the employed formalism.

These requirements for a conceptual model lead to two key questions about the modelling approach that should be followed:

- what to model, i.e. what aspects of the information system and the universe of discourse need to be captured in a particular conceptual schema?
- how to model, i.e. what particular concepts must a developer use for representing a slice of reality and what method should be followed for constructing an appropriate conceptual schema?

The answer to these two questions is the subject matter of the remainder of this chapter.

5.2 Conceptual Modelling Formalisms

A conceptual schema effectively carries knowledge about a universe of discourse. During the development lifecycle this knowledge is gradually built, modified and used, originally for purposes of communication between developers and users and subsequently for maintaining the integrity of a database. Obviously, the modelling of a slice of reality is entirely a matter of ontology and identification of the boundary of the universe of discourse (*Lee, 1986*). Therefore, it is not surprising that many different approaches are found in the literature in both fields of databases and artificial intelligence. In fact the task of modelling reality, as approached by the database and artificial intelligence communities, is regarded by many as the single most important integrating factor between the two fields (*Brodie et al, 1984*).

Central to the conceptual modelling process is a *formalism* (sometimes known as a conceptual modelling language) which uses a set of concepts and rules for building well-formed structures in a conceptual schema. The concerns of these formalisms are:

- the adequacy of the formalism's concepts for representing real world knowledge
- the efficient structuring of knowledge in conceptual structures in order to cope with problems of size and complexity
- the distinction between objects of the universe of discourse and the denotation of these objects
- the provision of graphical representations

- the handling of constraints about the concepts represented in the schema

- the provision of powerful abstraction mechanisms so that one can reason about the structures of the conceptual schema

- the mapping of structures from the conceptual level to the database level with the minimum of transformations between the two levels.

In mainstream information systems development conceptual modelling formalisms rely on the entity-based paradigm. Note that there are a number of research developments which use extensions on the basic entity-based approach, for example, TAXIS (*Mylopoulos et al, 1980*), CIAM (*Gustafsson et al, 1982*), ERAE (*Hagelstein, 1988*), CML (*Jarke et al, 1988*), RUBRIC (*van Assche et al, 1988*). A classification of data modelling approaches can be found in (Klein & Hirscheim, 1987).

Formalisms which follow the entity-based paradigm fall into one of two categories depending on their assumptions about the basic concepts of the universe of discourse, they are the entity-attribute-relationship (EAR) approaches and the binary-relationship (BR) approaches.

The EAR approaches assume that a universe of discourse can be described in terms of three basic primitives, the entity, the attribute and the relationship. An entity is regarded as the representation of an object in the universe of discourse about which information is required; attributes refer to the properties of an entity; and a relationship refers to an association between two entities. The major exponent of this philosophy is the E-R model (*Chen, 1976*).

The BR approaches avoid the *distinction* between attribute and relationship and in this sense it can be argued that the BR approach offers a greater level of abstraction than the EAR approach. A number of authors have criticised the EAR approach as being too close to the implementation level which in turn imposes conceptually irrelevant constraints on the formalism (*Kent, 1983*; *Nijssen et al, 1988*). A major exponent of the BR approach is the NIAM model (*Verheijen & van Bekkum, 1982*).

An important facet of all conceptual modelling formalisms is the distinction between *intensions* and *extensions*. The intension of a word is that part of meaning that follows from general principles. An example of an intension is "*employees work in departments*". The extension of a word is the set of all existing things to which the word applies. Extensions are normally large sets which cannot be observed in their entirety. For example the extension of *employee* is all the possible employees. In practical terms, for a specific information system one is interested, at any point in time, only in the extensions as represented in the database, but during analysis and design the developer will concentrate at the intensional level.

Intension and extension give rise to the syntactical (datalogical) and semantic (infological) views. The relationship between syntactic and semantic viewpoints is best represented by the well known meaning triangle (*Ogden & Richards, 1923*; *Sowa, 1984*) as shown in figure 5.3.

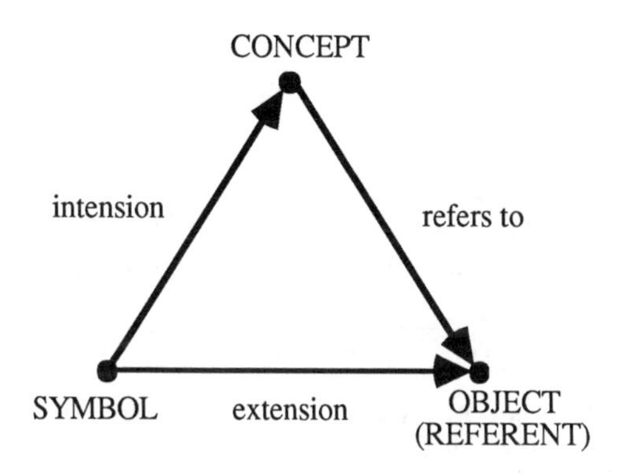

Figure 5.3: The Meaning Triangle

The peak of the triangle depicts the intension (concept, idea, thought or sense) and corresponds to the conceptual realm, whereas the base of the triangle corresponds to the objective world. Two types of concept can be distinguished: generic and individual. A generic concept has as a referent a set of objects (right hand corner) for example, *employee*. An individual concept has as a referent one particular object for example, *the employee J. Smith*. The left hand corner is the symbol (also known as word, sign, or data) and pertains to the datalogical point of view i.e. it refers to the representation of the object.

5.3 Basic Concepts

This section deals with the basic concepts found in entity-based formalisms. First, the concepts for the EAR approach are introduced (section 5.3.1) followed by a discussion of the concepts in the BR approach (section 5.3.2). Because of the greater conceptual expressiveness of the BR formalism, more emphasis has been placed in the coverage of a model which falls into this category, particularly in the way that the model may be applied in conceptual schema design. Naturally, there is some overlap between the two formalisms and the astute reader, having studied the material of subsection 5.3.1, will have little difficulty in also applying the E-R model to the examples presented in section 5.3.2.

5.3.1 The Entity-Relationship Model

The entity-relationship (E-R) model has been widely applied and forms part of many contemporary information systems development methods. The basic concepts for this model are those of entity, attribute and relationship.

The Entity Type

An entity is defined as 'something about which information is recorded'. For example the following may be regarded as entities:

<1245, Smith, Accounts, 20,000>
<3456, Jones, Marketing, 25,000>

<D1, Accounts, Brown>
<D2, Marketing, Jones>

The entities in the first set correspond to information kept about employees whereas the entities in the second set correspond to information about departments. Obviously, when developing a conceptual schema there is a need to abstract from individual occurrences and refer to a generic concept (i.e. deal with objects at the intension level). The concept which satisfies this requirement is that of the *entity type*. In the above example there are two entity types, EMPLOYEE and DEPARTMENT. An entity type may be defined as:

$$E_i = \{e_i \mid P_i\}$$

that is, all entities e in an entity set E have the same properties p. Examples of entity types in the hotel case study are HOTEL, ROOM, RESORT, BILL, CHARGE-ITEM. There are many entities in an organisation and it is the responsibility of an analyst to select the entity types which are relevant to the information system under development. Identifying entity types is very often the source of errors, since the same object may have a different meaning to different people. Reconciliation can only be achieved if the analyst develops an understanding of information structures and verifies this against the views of organisation personnel.

The Attribute Type

Every entity has properties which are expressed in terms of *attribute-value* pairs.

Each attribute is referenced by an attribute name and corresponding to this name there will be one or more values. For example, the attribute CUSTOMER-NAME may have *J. Smith* as its value and will correspond to a single entity occurrence i.e. *the customer whose name is J. Smith.* It is possible of course, although not that common, that an attribute may have multiple values. An example of this is HOTEL-TEL-NO where a hotel has more than one telephone number.

For the same reasons as entities, attributes are grouped into attribute types. An *attribute type* is an aggregation of allowable values (drawn from a domain) for a property of an entity type. For example, ROOM-NUMBER is an attribute type whose values are drawn from the domain of non-negative numbers.

In some cases it is convenient to define an entity in terms of compound attributes, i.e. they can themselves be regarded as entities with a set of attributes. For example, the entity invoice might have attributes, *invoice, number, product, unit cost, number, price, date. Date* could be regarded as a compound attribute and it may be necessary to access the individual parts such as the month. To reduce this information to a standard form, the attributes in an entity should be made elementary, i.e. non-divisible, by expanding any compound attribute.

The need often arises to address individual entities. This can be achieved by the use of an *entity identifier*. An entity identifier is one or more attribute types whose value uniquely determines an entity occurrence. For example, the entity type ROOM may have as its identifier the attribute type ROOM-NO. This means that each value of ROOM-NO is unique, thus enabling the exact identification of a ROOM entity occurrence.

It is possible that for some entity types there may not be a natural identifier. If this happens an analyst should consider an artificial identifier by choosing the one or more of the existing attributes or even introducing new ones.

Figure 5.4 shows an example of an entity type CUSTOMER having three attribute types CUSTOMER-ACCOUNT, CUSTOMER-NAME and RESERVATION and for each one of these a domain of values. The example also shows that for the entity e_1 its attributes take the values (A443, N.White, R1234) with A443 being its identifier.

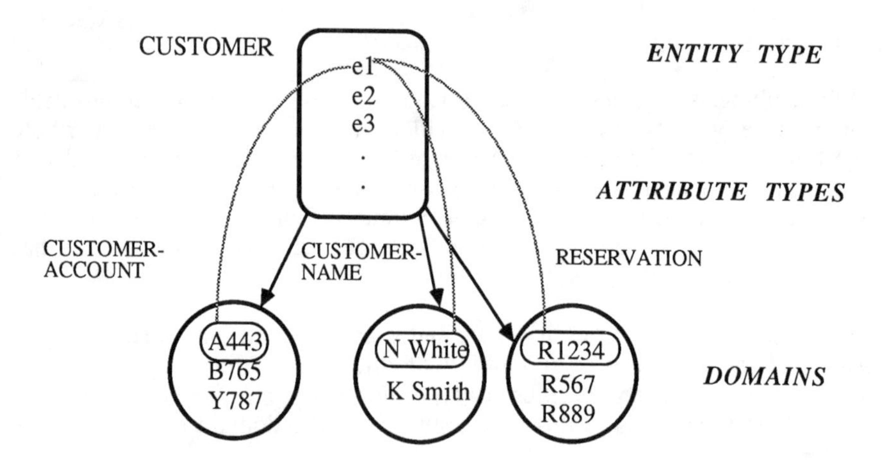

CUSTOMER (CUSTOMER-ACCOUNT, CUSTOMER-NAME, RESERVATION)

Figure 5.4: Entity Types, Attribute Types and Values

The Relationship Type

An association between two or more entity types is known as a *relationship*. Figure 5.5 shows instances of entity types HOTEL and RESORT and instances of relationships between these entities. An aggregation of relationships involving the same entities gives rise to a relationship type and in general this is defined as:

$$R = \{<e_1,....e_n> \mid e_1 \in R_1.E_1,......e_n \in R_n.E_n\}$$

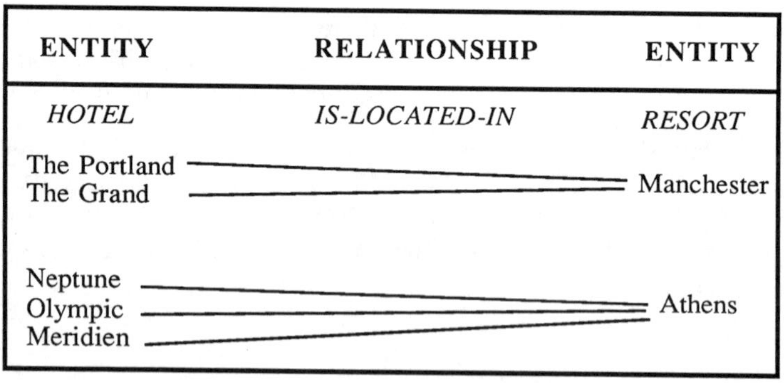

ENTITY	RELATIONSHIP	ENTITY
HOTEL	*IS-LOCATED-IN*	*RESORT*
The Portland The Grand		Manchester
Neptune Olympic Meridien		Athens

Figure 5.5: Entity and Relationship Occurrences

In reasoning about some part of the universe of discourse, a developer will refer to relationship types rather than individual instances. Consider for example the following statement "a customer makes a request for a hotel for a specific period of time". In this statement four entity types are identified, CUSTOMER, REQUEST, HOTEL-TYPE and PERIOD. The semantics of this statement can be defined by relating these four entity types as shown in figure 5.6.

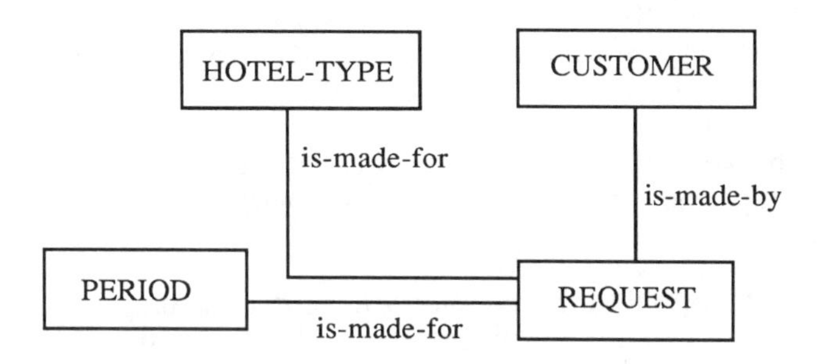

Figure 5.6: Relationships Between Entity Types

The decision as to the most appropriate way of modelling relationships between different objects is very subjective and depends primarily on the understanding of the business environment by the system developer.

In order to identify a particular relationship between two entity occurrences, a *relationship identifier* is required. A relationship identifier is the composite of the identifiers of the entities that participate in the relationship.

Relationships have properties which an analyst may wish to record. Relationship attributes are often used in enhancing the understanding of the semantics of data, although it should be noted that a number of development methods do not consider relationship attributes as part of their data analysis approach.

There are three ways in which two entity types (or more generally objects) can be related, a simple association, a complex association or a conditional association.

Simple Association. A simple association (type 1) between A and B exists when any value A uniquely identifies exactly one B.

Complex Association. A complex association (type M) between A and B exists when each A can be associated with any number of B.

Conditional Association. A conditional association (type C) between A and B exists when any A is associated with either one or none of B.

A useful concept in modelling relationships between different objects is that of mapping. A mapping between A and B is an association from A to B and its reverse.

The relationship between A and B is specified by giving the type of the mappings A→ B and B→ A so that there are the following possibilities (1:1), (1:M), (1:C), (M:1), (M:M), (M:C), (C:1), (C:M), (C:C). The ones of most interest are the (1:1), the (1:M), (M:1) and (M:M).

A mapping between two entity types is often referred to as the *cardinality* of the relationship. The mappings between two objects A and B are interpreted as follows.

One-to-One Mapping (1:1): For any A there may be only one member of B **and** for B there is only one member of A associated with it.

One-to-Many Mapping (1:m): For any A there may be many members of B **and** for any B there is only one member of A associated with it.

Many-to-One Mapping (m:1). For any A there may be only one member of B **and** for any B there may be many members of A associated with it.

Many-to-Many Mapping (m:n). For any A there may be many members of B **and** for any B there are many members of A associated with it.

The use of relationship cardinality as well as more complex cases of relationship types involving more than two entity types (or even entities of the same type) are elaborated in section 5.3.2.

The E-R Diagram

A recurring theme in most information system development techniques is the use of diagrams for depicting aspects of specification. The E-R model makes use of a diagramming technique known as the E-R diagram which can be used for representing entity types, attribute types, relationship types and relationship cardinalities. An example of the use of the E-R diagram is shown in figure 5.7. The notation used here is the basic notation used when the E-R model was first introduced (*Chen, 1976*). However, it should be noted that many development methods have adopted different notations whilst retaining the semantics of the E-R model itself.

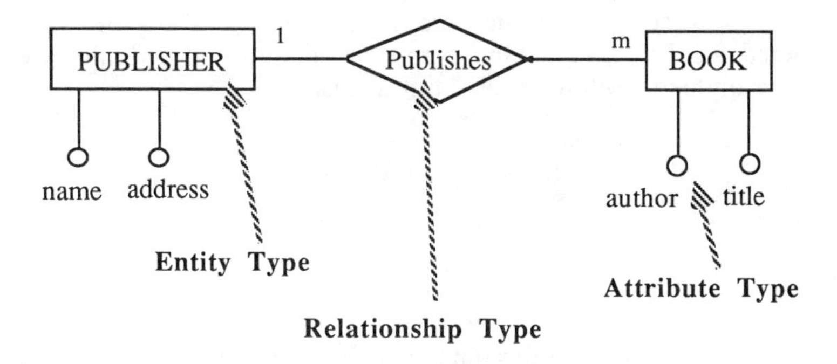

Figure 5.7: An E-R Diagram

The E-R diagram of figure 5.7 shows all basic concepts of an E-R model including the relationship cardinality (1:M, i.e. a publisher publishes many books, but a book is published by only one publisher).

5.3.2 The Fact-Based Model

The fact-based model, as discussed already, falls in the category of the Binary-Relationship (BR) approach whose major difference from the EAR approach is that no distinction is made between attributes and relationships.

In the fact-based model a conceptual schema is viewed as a set of individual *elementary facts*. An elementary fact is a true elementary proposition for example, consider the following elementary propositions:

1a. "Carol studies Systems Analysis"
1b. "Systems Analysis is studied by Carol"

2a. "John works for Accounts"
2b. "Accounts employ John"

3a. "Carol works for Data Processing"
3b. "Data Processing employs Carol".

The basic concepts found in the fact-based model are described next.

The Label Type

References to individual things in the universe of discourse is made through labels. Examples of labels taken from the above set of elementary propositions are *Carol*, *Systems Analysis*, *John*, *Accounts*. A developer needs to abstract from individual occurrences and deal with things of the universe of discourse at the intensional level. Therefore, labels are abstracted to *label types*. A label type is an aggregate of several labels. In the above example the following label types exist:

Name: {Carol, John}
Department Name: {Accounts, Data Processing}
Subject Name: {Systems Analysis}

The Entity Type

A thing of the universe of discourse which is referenced by a label is known as an entity. An entity type is an aggregate of entities for example, PERSON, DEPARTMENT, SUBJECT. Therefore, entities and labels clarify the semantics of a particular situation. The elementary propositions can now be defined as:

1a. Person with name 'Carol'
studies
Subject with subject name 'Systems Analysis'

2a. Person with name 'John'
works for
Department with department name 'Accounts'

The Relationship Type- Fact and Reference Types

A relationship is an association between two or more things in the universe of discourse. A set of relationships between the same things gives rise to a relationship type.

A relationship type between an entity type and a label type is called a *reference type*. An example of a reference type is 'Person has Name' where PERSON is an entity type and Name is a label type.

A relationship type between two entity types is called a *fact type*. An example is, 'Person studies Subject' indicating that both PERSON and SUBJECT are entity types.

Each relationship type (reference or fact) represents one deep structure for which there may be two or more surface structures (depending on the number of things that the relationship type associates). Each surface structure represents the *role* (also called *sentence predicate*) that an entity type or a label type plays in the relationship. For example, 1(a) and 1(b) below together represent a single, deep structure i.e. a relationship type between two entity types.

1a. Employees work in Departments
1b. Departments employ Employees

The sentence predicate 1(a) represents the role that the entity type EMPLOYEE plays in its association with the entity type DEPARTMENT. The sentence predicate 1(b) represents the inverse.

Fact types are not restricted to being *binary* i.e. involving only two entity types. In general it is possible to have n-ary fact types where n is the number of associated entity types. Figure 5.8 shows a ternary fact type. Each row in the table of labels represents an instance of a fact whereas the whole table would represent the population of the ternary fact type.

Supplier	Product	Price	**Entity Types**
Supplier_name	Product_no.	Amount	**Label Types**
supplies	*supplied*	*priced_at*	**Roles**
J. Smith	P123	30.00	**Labels**

Figure 5.8: A Heterogeneous Ternary Fact Type

Fact types may be established between different entity types (heterogeneous fact type) or between instances of the same entity type (homogeneous fact type). Thus, the fact type in figure 5.8 is an example of a heterogeneous fact type. Figure 5.9 shows two examples of homogeneous fact types; in the first case the fact type is symmetric i.e. both roles are the same; in the second case the fact type is asymmetric i.e. the roles are different.

It is often necessary or convenient to regard a fact type as an entity type so that it can be related to another entity or label type. The process whereby a fact is represented within another fact is known as *nominalisation* and a fact type which is viewed as an entity type is known as an *objectified* fact type.

As an example consider the case where one wishes to model the situation where 'students study subjects on a part time basis' and in addition model 'the number of hours that each student studies a subject'. A tabular notation of this is shown in figure 5.10. In this case the fact type between student and subject is objectified to be the entity type assignment.

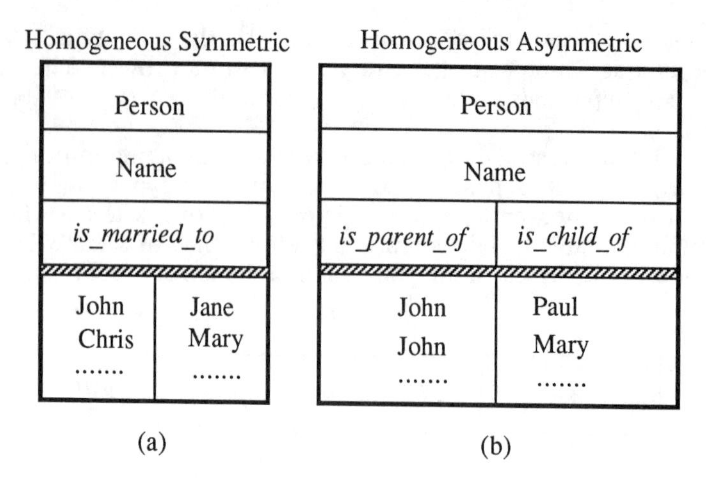

Figure 5.9: Homogeneous Fact Types

As in the case of the E-R model, the specification of relationship types also involves their cardinality. As discussed in section 5.3.1 a relationship cardinality is concerned with the mapping between the involved entity types. For a fuller discussion on this topic the reader is advised to refer to section 5.3.1.

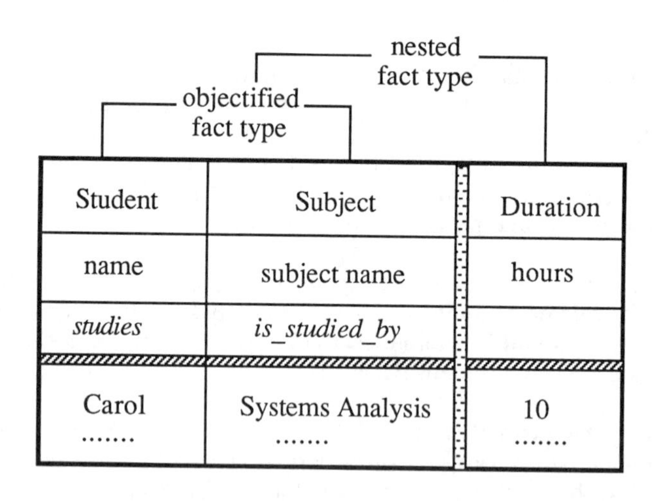

Figure 5.10: Objectified Fact Type

The Information Diagram

Label types, entity types and relationship types (fact and reference) can be combined in a diagram known as the information diagram. The notation used in this chapter is summarised in figure 5.11.

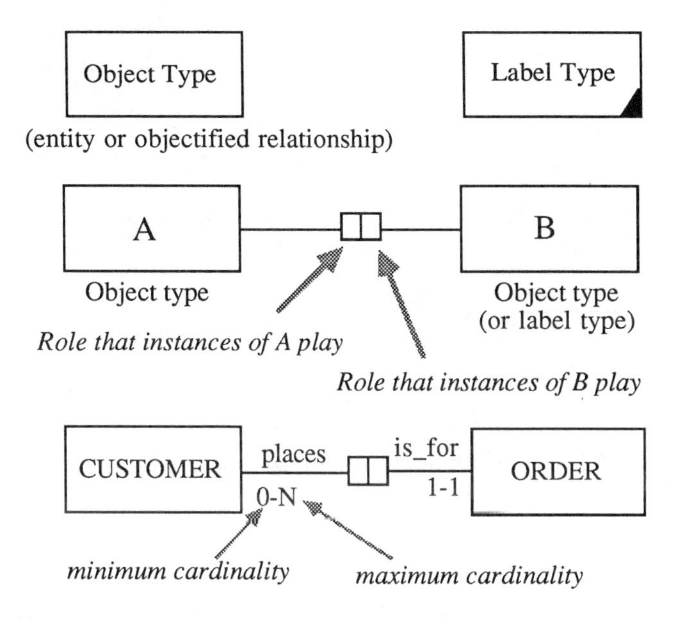

Figure 5.11: Primitives of the Information Diagram

An example of the use of the information diagram is shown in figure 5.12.

A relationship cardinality in an information diagram represents the minimum and maximum number of roles that each entity may participate in the relationship. The forms of cardinality considered are those of: zero or one (0-1); zero, one or more (0-N); one and only one (1-1); and one or more (1-N).

As an example consider the relationship type between customer and order in the diagram of figure 5.12 which is interpreted as "a customer may place zero one or many orders and an order is placed by one and only one customer".

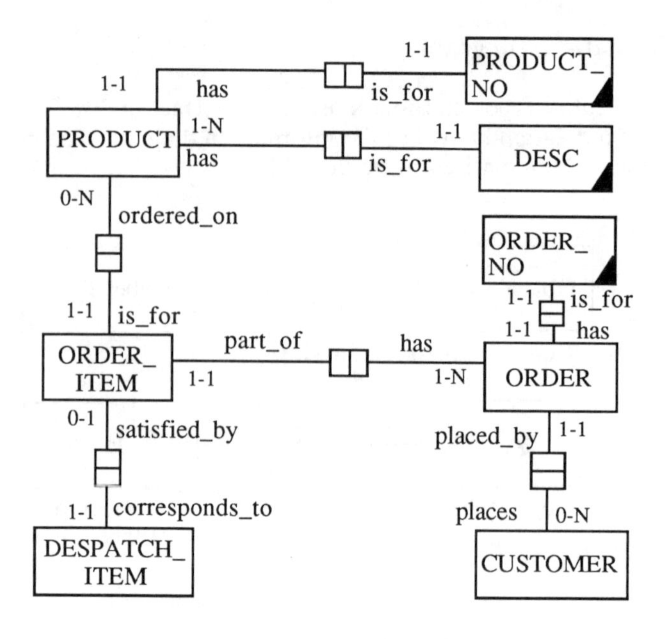

Figure 5.12: Example of an Information Diagram

Entity Subtypes

In developing information systems it is often natural and convenient to start with some generalised concepts and, as a clearer understanding is obtained, to develop more specialized models. This concept has already been discussed in chapter 4 in the context of process decomposition in the data flow model.

In conceptual data modelling the concept of specialization is accommodated by *entity subtyping*.

An entity type E2 is a subtype of another entity type E1, if E2 has all the properties of E1. Any additional properties of E2 are not shared by E1. In this way entity type E2 is given a narrower definition than E1. In the information diagram, entity subtyping is indicated by a special relationship, known as the *is_a* relationship, denoted as an arrow from the subtype to supertype.

In the diagram of figure 5.13 LECTURER is a subtype of EMPLOYEE. This means that some members of EMPLOYEE will belong to the special case of being a member of LECTURER.

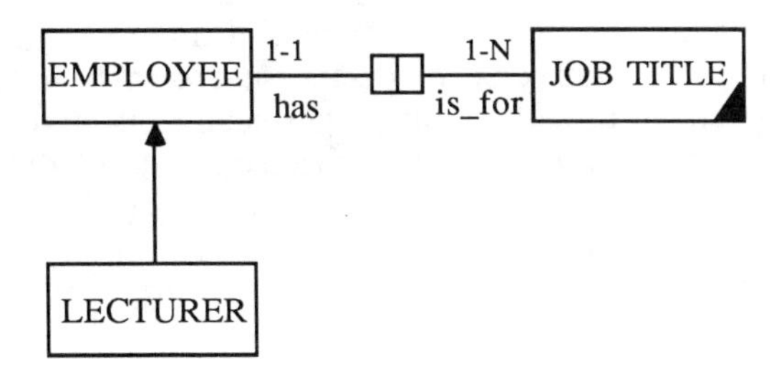

Figure 5.13: Entity Subtype

In general, a specialization is determined be evaluating a predicate which indicates whether an entity belongs to the specific subtype. In the example of figure 5.13 the classifying predicate of LECTURER is:

<JOB TITLE of EMPLOYEE = 'LECTURER'>

An entity subtype may itself by specialized into subtypes. Figure 5.14 shows an example of such a case.

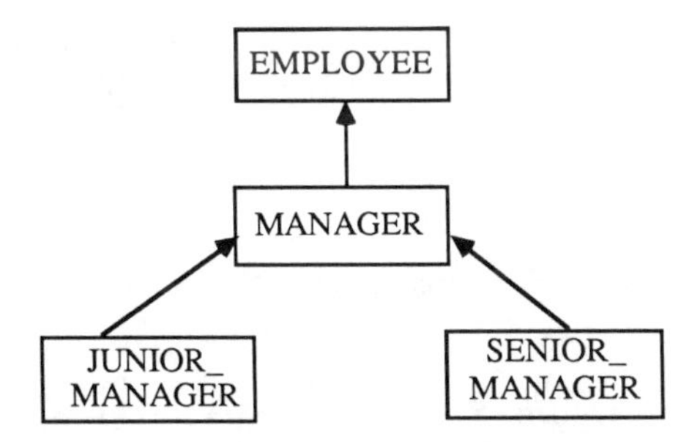

Figure 5.14: A Subtyping Hierarchy

Modelling Time

The need to deal with temporal aspects of information systems has been investigated in databases from the perspective of historical information (*Wiederhold et al, 1975*) and in artificial intelligence from the perspective of planning (*Allen, 1983*; *McDermott, 1982*; *Dean, 1987*). For details on these approaches the interested reader can refer to the surveys by Bolour et al (*1982*) and Jardine & Matzov (*1986*). For a discussion on temporal logics and their application to databases and artificial intelligence, the collection of papers edited by Galton (*1987*) is very useful material.

In conceptual data modelling, the treatment of time is of crucial importance if the semantics of a universe of discourse are to be clearly defined. A number of conceptual modelling languages such as RML (*Greenspan, 1984*) and ERAE (*Haggelstein, 1988*) offer advanced features of time modelling. However, the discussion here is confined to the type of models discussed so far. To this end, time is modelled either indirectly, via the use of relationship cardinalities, or directly by representing a required facet of time as an entity type.

The relationship cardinality can be used to indicate either a *snapshot* or a *historical* situation. For example, consider the case where employees in a company can work in many different departments while they are employed by the company, but at any point in time an employee can work for one department only. This situation can be modelled in two different ways depending on the requirements of the application. The alternatives are shown in figure 5.15.

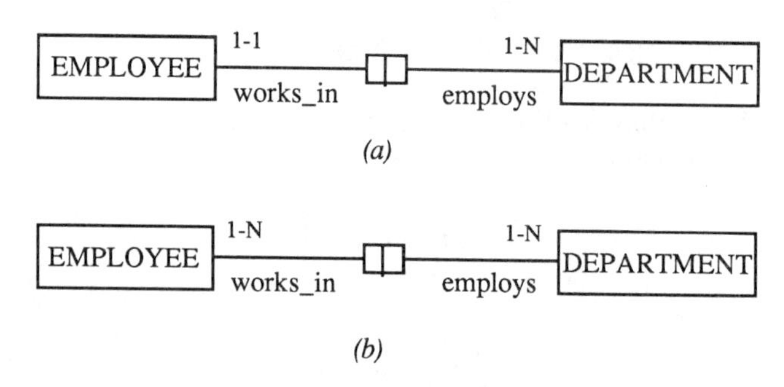

Figure 5.15: Snapshot and Historical Situations

The modelling in figure 5.15(a) indicates that one is interested in information which relates to a specific point in time (snapshot). This is because what is modelled is that "an employee can work for one and only one department". The modelling in figure 5.15(b) deals with historical information, as can be seen by the relationship cardinality which states that "an employee may work in one or many departments". A further use of the cardinality concept in modelling time is the use of the zero cardinality. Figure 5.16 shows a different situation to that modelled in figure 5.15.

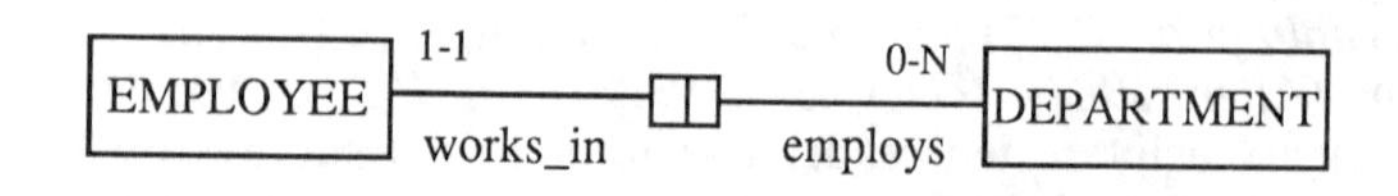

Figure 5.16: The Use of Zero Cardinality in Time Modelling

What is modelled in figure 5.16 is the case when a department is created and has not had any employees allocated for working for that department.

A different approach to modelling time is the use of an entity type to *explicitly* represent a required function of time. For example, consider the case where patients attend appointments in a hospital and for each appointment there is a start time. Figure 5.17 shows how this situation may be modelled.

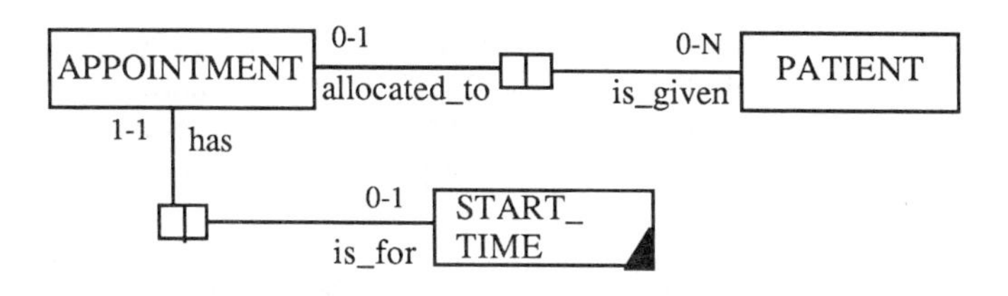

Figure 5.17: Explicit Representation of Time

5.4 The Conceptual Modelling Activities

The conceptual modelling process is characterised by two key activities, view modelling and view integration, as shown in figure 5.1. The activity of view modelling is concerned with the derivation of schemas for individual groups of users or application domains. The activity of view integration attempts to reconcile any differences between individual views and synthesize all these views into a single, global conceptual schema.

However, conceptual data modelling cannot be carried in a vacuum, independent of any consideration about the dynamic behaviour of the universe of discourse. There is a need to cross-reference the process modelling phase in order to ensure that:

- from the process perspective all required data has been identified
- from the data perspective all processing requirements can be accommodated by the conceptual schema and database.

The main activities involved in conceptual data modelling are as follows.

Cross Referencing with Process Model. This stage ensures that the data view and the process view support each other and are consistent across the spectrum of applications. In practice, there are two possible way for achieving this. One way is to relate data to processes using the *data-process matrix*. This matrix can be used at various levels of process decomposition. At the lowest level, it is possible to define the type of primitive operations (insertion, deletion, accessing) each process may invoke upon data. Another way of cross referencing the two modelling orientations is to model the *life history* of each entity type. In this case, the behaviour of an entity type in reaction to events which may take place throughout the expected life of the entity type is modelled.

View Modelling. This activity essentially involves many of the issues discussed so far in this chapter. A conceptual schema is developed, using one of the entity-based approaches, for each application area. The need for developing individual conceptual schemas is the same is that for process decomposition.

View Integration. Once all areas under examination have been covered, each data model can be integrated into one *global corporate conceptual schema* to serve as a single reference point for all processes.

In many contemporary development methods a technique known as *data normalisation* is included in the data analysis phase. This technique is covered in the chapter on data design (chapter 8), since it is the authors' belief that, within the context of conceptual data modelling as presented in this chapter, data normalisation is a technique more appropriate to data design.

5.4.1 Cross Referencing Data to Processes

The Data-Process Matrix

As previously mentioned, conceptual modelling cannot be effectively carried out without considering process modelling. Consideration of a system's processes will enhance the developer's understanding of that system's data. For example, in the hotel case study analysis of the *request for room* function will reveal all entity types and their interrelationships as they pertain to that particular function. Therefore, the modelling of organisational processes and data are activities which should be carried out in parallel, so that knowledge from one type of model is used to construct and refine the other.

The result of relating data to processes can be documented as a data-process matrix as shown in figure 5.18.

Data \ Process	Service Request	Make Reservation	Invoicing	Checking in	Check-out	Billing
Customer		X	X			X
Request	X					
Reservation		X		X		
Hotel-type	X					
Room		X		X	X	
Bill						X
Bill-item						X
Period	X					
Tariff			X			
Resort	X					
Hotel		X		X	X	

Figure 5.18: Relationships Between Data and Processes

The horizontal axis of this matrix shows the processes which are involved in the hotel organisation. The vertical axis records all the different types of data that exist within this organisation. This type of matrix can also be used to annotate the usage of each entity by each process in terms of the basic data operations of *read*, *write*, *delete* and *update*. In this case instead of an X a matrix cell will have one of R, W, D or U to indicate the appropriate data operation.

Entity Life History Analysis

Entity life history analysis is concerned with documenting details of the behaviour of entity types in reaction to events which take place within the system under consideration.

The objective of this technique is to identify and document the *states* of an entity type throughout its lifecycle. Like many other techniques used by contemporary development methods, modelling of entity life histories can be specified by using graphical means. There are a number of different notations available to an analyst, for example (*Jackson, 1975*; *Rock-Evans*; *MacDonald, 1986*).

For each entity type, all the events which may affect an occurrence of the entity type during its lifetime are documented. The only events which are considered are those events which trigger some action to take place in the system. For example, the event *order arrives* triggers the elementary function *process order* which can be decomposed into a number of system actions such as *check creditworthiness*, *check stock*, *raise invoice* etc.

The entity type ORDER may be involved in different business functions during its life depending on whether for example there is enough stock, the customer's credit is acceptable etc. This entity type therefore can exist in different states and this existence is guided by an event taking place.

The entity life history analysis is another way of relating process models to data models. This activity imposes an almost formal way of examining all processes which use a particular entity type. The advantage of the technique at the analysis stage is to ensure that all possible events associated with an entity type have been considered. Furthermore, the knowledge of what processes are valid in each state of the entity type can be used during transaction system design.

5.4.2 View Modelling

View modelling is concerned with constructing *local* conceptual schemas by considering the data requirements of individual application areas. The term local is used to refer to the fact that each conceptual schema would be responsible for a 'community of users with a common function within the organisation'. In the case of large applications where there are several user groups each one with a particular view of an application and different needs of the database this approach is more feasible rather than attempting to design a conceptual schema for the whole organisation in one step.

Taking the hotel case study as an example, two functional areas are those of *request* and *billing*. The corresponding local conceptual schemas for these two areas are shown in figures 5.19 and 5.20 respectively.

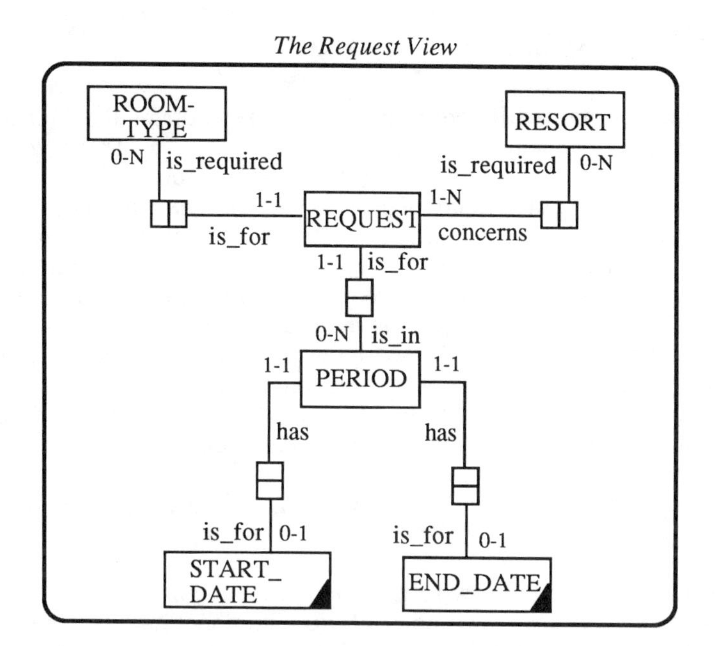

Figure 5.19: The Request Local View

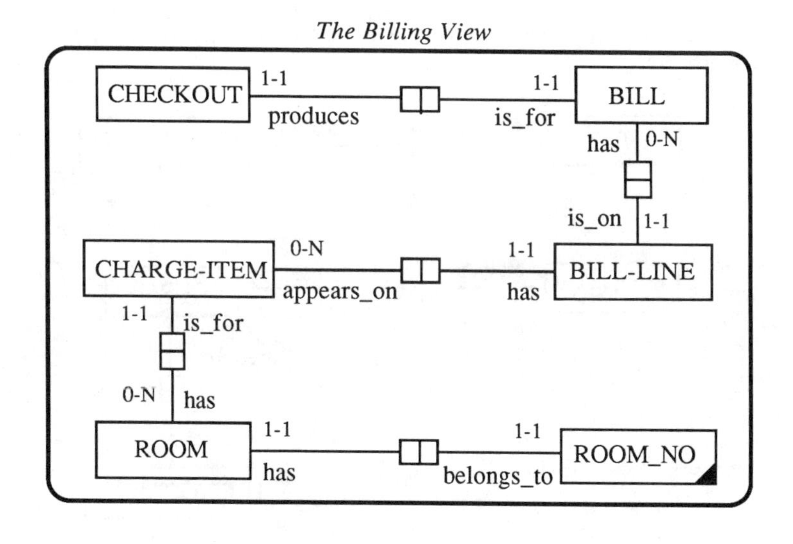

Figure 5.20: The Billing Local View

5.4.3 View Integration

The existence of many different user schemas necessitates their integration into a single schema against which a database will operate. This gives rise to the second major activity in conceptual schema design namely view integration. The main objective of *view integration* is to resolve any semantic and structural differences between the various user views and to synthesize these into a global conceptual schema. Problems of inconsistency arise mainly because user schemas are developed by different designers and because the viewpoint of each schema is influenced by the requirements and particular semantic interpretations of the user application domain to which the schema relates. The source of these inconsistencies may be in the use of a different linguistic approach in naming various objects or from the adoption of different views in modelling the same object.

The main causes for schema diversity (*Batini & Lenzerini, 1984*) are as follows.

Different Designer Viewpoints. In the design process, different designers adopt their own point of view in modelling the same objects in the application domain. Figure 5.21 shows an example of difference in perception in the design of the same concept. Both schemas represent information about students and their department. But, in schema 1 the association between STUDENT and DEPARTMENT is a direct one, as represented by the relationship type *registers*, whereas in schema 2 the same association is perceived through the entity type COURSE and represented by the two relationship types *attends* and *offered_by*.

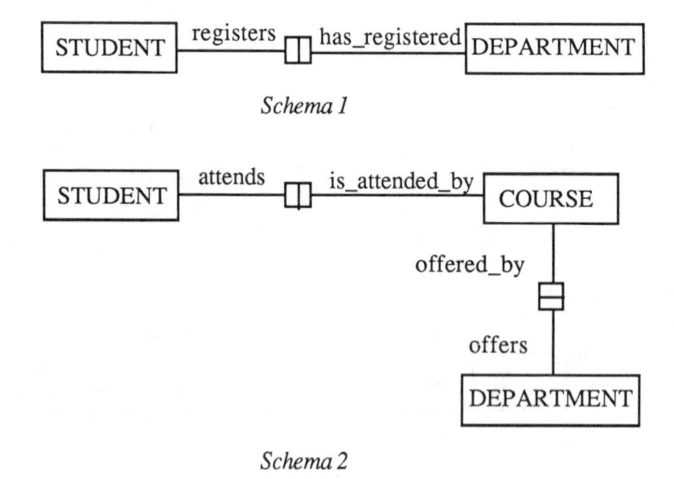

Figure 5.21 : Different Design Viewpoints

Equivalence Among Constructs of the Schemas. In constructing a schema, several combinations of constructs can model the same application domain equivalently. A frequently occurring situation is where relationship types and label types are used interchangeably. An example is given in figure 5.22, in which the entity type PUBLISHER in schema 1 is associated with the entity type BOOK by the relationship type *publishing*, whereas in schema 2 PUBLISHER is represented as label type of the entity type BOOK.

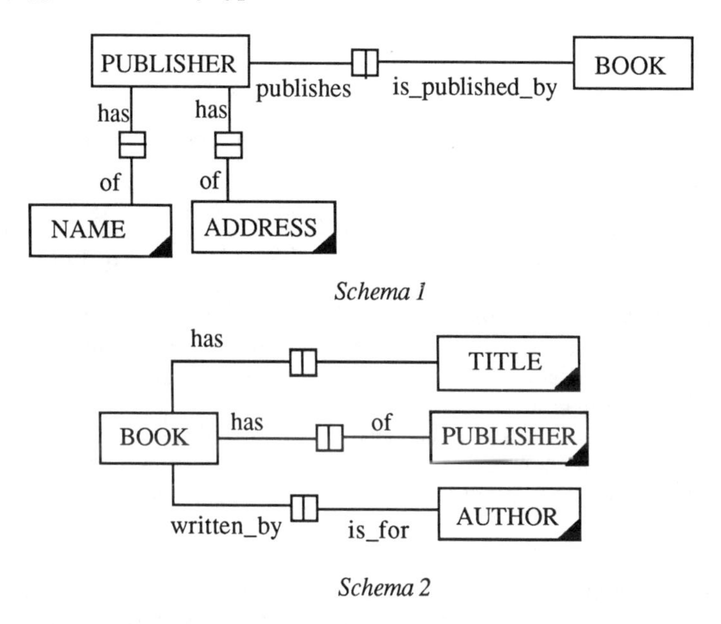

Figure 5.22 : Equivalent Constructs

Incompatible Design Specifications. Two or more schemas may be incompatible because of incompatible design decisions about names, types, integrity constraints etc., of corresponding constructs. In figure 5.23, schema 1 shows the assumption that an EMPLOYEE may not necessarily be assigned to a PROJECT, since the cardinality constraint 0-N has been specified (optionality), while in schema 2, it has been assumed that an EMPLOYEE must be assigned to at least one PROJECT.

Component schemas may also contain concepts that are not the same but are related by some semantic relationship, thus giving rise to interschema relationships. In order to perform integration it is important not only to resolve the various inconsistencies between the common concepts, but also to discover all the interschema relationships, between different concepts.

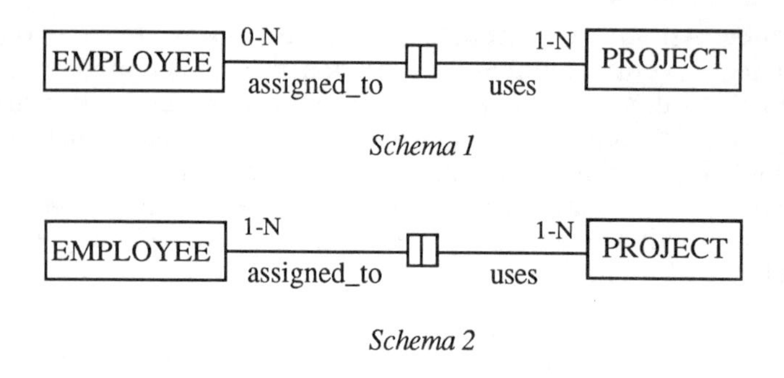

Figure 5.23: Incompatible Design Specifications

Several methods for view integration have been proposed (*Al Fedaghi & Scheuerman, 1981*; *Batini & Lenzerini, 1984*; *Casanova & Vidal, 1983*; *El Masri et al, 1987*; *Kahn, 1979*; *Navathe & Gadgil, 1982*; *Teorey & Fry, 1982*; *Wiederhold & El Masri, 1979*; *Yao et al, 1982*). Each method tackles the problem with different strategies and in the context of different modelling formalisms. However, common to all these methods is a set of activities whose goal is to find all common parts of the various schemas, discover all conflicts that exist between them and resolve them (so that all user schemas are compatible for merging into a global schema), define a strategy for the view integration process and finally carry out the integration process.

The activities involved in conflict detection and resolution involve *semantic checking* and *schema transformation*. Incompatibilities are determined by carrying out conflict analysis and interschema relationship analysis. The goal of conflict analysis is to discover all the differences that exist in representing the same classes of objects in different schemas. The process is concerned specifically with *naming* (i.e. homonyms and synonyms) and *structural* (i.e. type, dependency, identifier and behavioural) conflicts. The activity of interschema relationship analysis is concerned with discovering the different concepts in different schemas that are mutually related by some semantic relationship.

The integration process includes all the activities involved in the actual merging of the schemas. These activities are, the *choice of an integration procedure* and the *merging of entities and relationships*. The nature of these activities depends entirely on the choice of method chosen.

An integrated conceptual schema for the hotel case study is presented in figure 5.24.

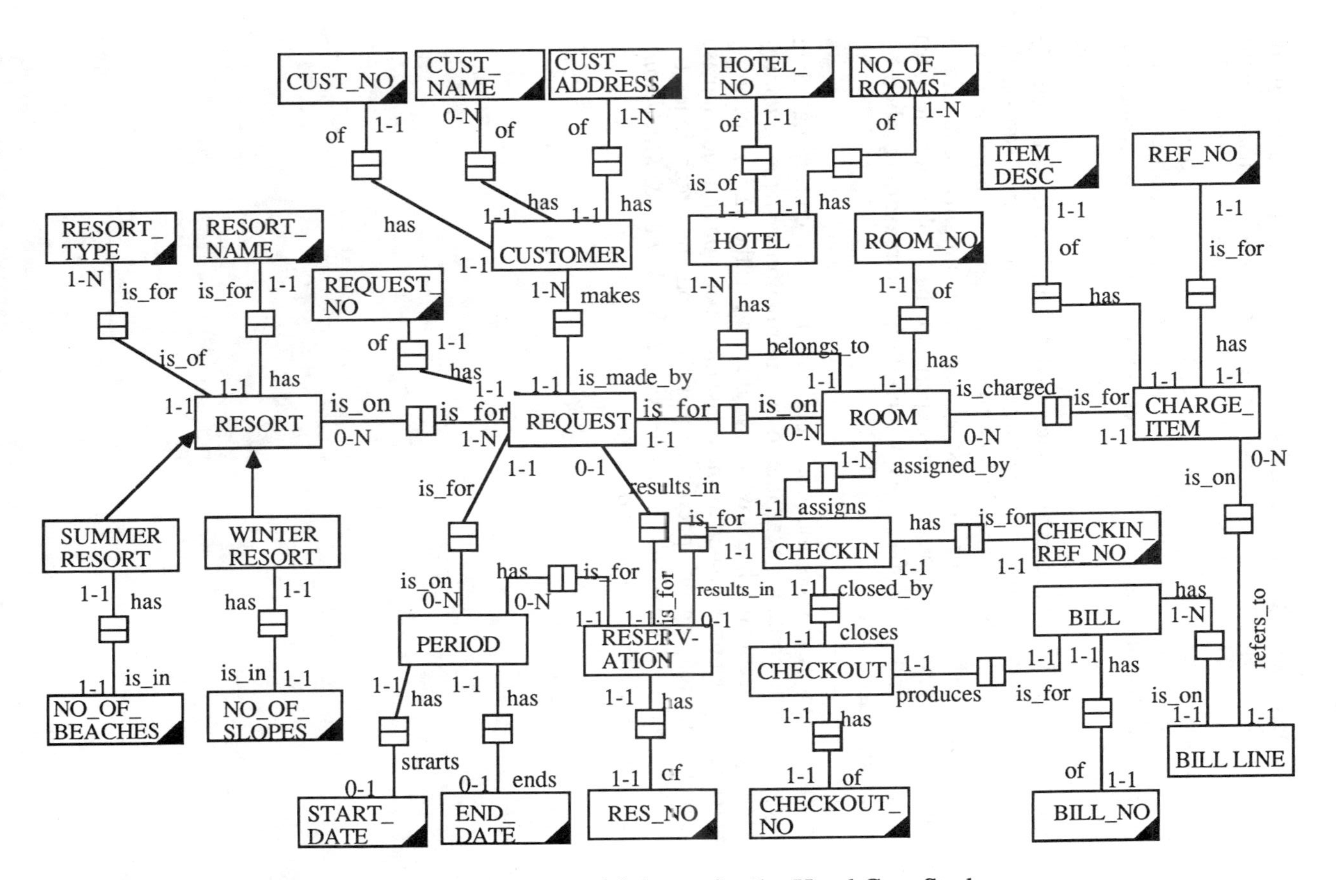

Figure 5.24: A Global Conceptual Schema for the Hotel Case Study

5.5 Summary

Conceptual data modelling is concerned with analysing the static elements of a universe of discourse and deriving a schema which is precise, unambiguous and non-redundant. This is achieved by carrying out a set of activities which involve the cross referencing of the data specification to the process specification, developing a conceptual schema for each application and integrating all different user schemas (views) into a single global conceptual schema.

An important factor in the success of conceptual data modelling is the choice of an appropriate conceptual modelling formalism. Such a formalism should provide structures which are independent of any implementation considerations, offer powerful abstraction mechanisms, encourage communication between end-users (the experts in the application domain) and developers and offer facilities which help in reasoning about a conceptual schema.

This chapter has discussed the concepts involved in conceptual data modelling and the techniques available to an information systems developer for designing a conceptual schema which would serve as the basis of defining all data types in a database as well as the allowable operations on the database itself. For further reading on specific conceptual modelling techniques the reader is referred to (*Tsichritzis & Lochovsky, 1983*; *Vetter & Maddison, 1981*; *Rock-Evans, 1981*; *Howe, 1983*).

Chapter 6

Automated System Design

Previous chapters have described the concepts and techniques which enable analysts to capture and embody facts in requirements specifications. The process of producing a requirements specification will have involved the participation of many individuals, each contributing varying amounts of knowledge about particular aspects of the system that is under analysis. At the same time, the system developer will have coordinated and shaped the information that he is given to reflect a specification of the *desired* system.

Once this point has been reached, the developer must begin to make the transition from what was referred to earlier as the expansion phase, to the contraction phase. In other words, the outputs from the process and data modelling activities, which expressed wishes and desires, must now be transformed into reality. This chapter sets the scene for this process of transformation. It begins with a discussion on the process of ensuring the correctness of a specification, followed by an examination of the issues that start to shape the emerging system.

6.1 Review of the Objectives of Analysis

The objectives of the analysis phase in the development of software is to identify the functions and characteristics of a required system. This activity may be based upon the analysis of an existing system, either manual or computer-based or it may be based purely upon the notions held by one or more individuals. The important point to remember is that however the system works at the moment, it is going to be different in the future; a manual system may become computer-based or an existing computer-based system may be amended.

The relevance of this notion of change from one system to another lies in the fact that analysis is concerned with finding out what activities a system must perform. How these activities are currently performed is of no direct interest to the analyst, although because the early stages of analysis are based upon techniques such as observation, questioning and interviewing, the analyst will of necessity discover how activities may currently be performed. However, it is almost certain that such information will initially become embodied in a requirements specification. Thus much of the transition from the analysis phase to the design phase is based upon ensuring that a requirements specification contains only the logical requirements of a system (*the what*) and nothing of the physical characteristics of the present system (*the how*).

But the transition from analysis to design also has another important aspect, that of validation of the requirements specification. It must be remembered that the requirements specification will state the requirements of a new system. These requirements are largely determined by those who are commissioning a system (*the acceptor*) and the transition from analysis to design represents the last, realistic point at which users can have any significant influence on the future system. Therefore prior to the transition to design, users must ensure that a requirements specification embodies their needs. Of course, user validation should take place throughout the analysis process, but it is incorporated at this point in this book because the transition from analysis to design represents the last opportunity for a user to embody his views and requirements in a specification.

Once a logical specification has been produced and validated, the system developer may start to address issues of how to build the system.

6.2 Ensuring a Logical Model

As previously noted, no analyst will attempt to specify a set of data flow diagrams that are totally devoid of physical considerations. Often physical considerations are of use and play an important role in the early stages of analysis. Typically such considerations help to get the process of analysis started and help users to identify with the documentation of the system. However, ultimately such considerations must be removed. Figure 6.1, for example, shows a data flow diagram which contains physical considerations, some of which are obvious and others which are more subtle.

The physical considerations that are embodied in requirements specifications may arise from a number of causes. These causes are as follows.

Procedural. The most obvious type of physical consideration which finds its way into a data flow diagram are those concerned with procedure. Typical examples involve the use of specific staff names, document or form names, organisation, department and section names and existing manual or automated systems.

In figure 6.1, the *delivery system* and *code-10-slip* are both examples of procedural characteristics which should be renamed since the names portray very little to a reader who is not an active participant in the system.

Historical. Historical considerations relate to those objects (data flows, processes and files) which have been lumped together for *convenience*.

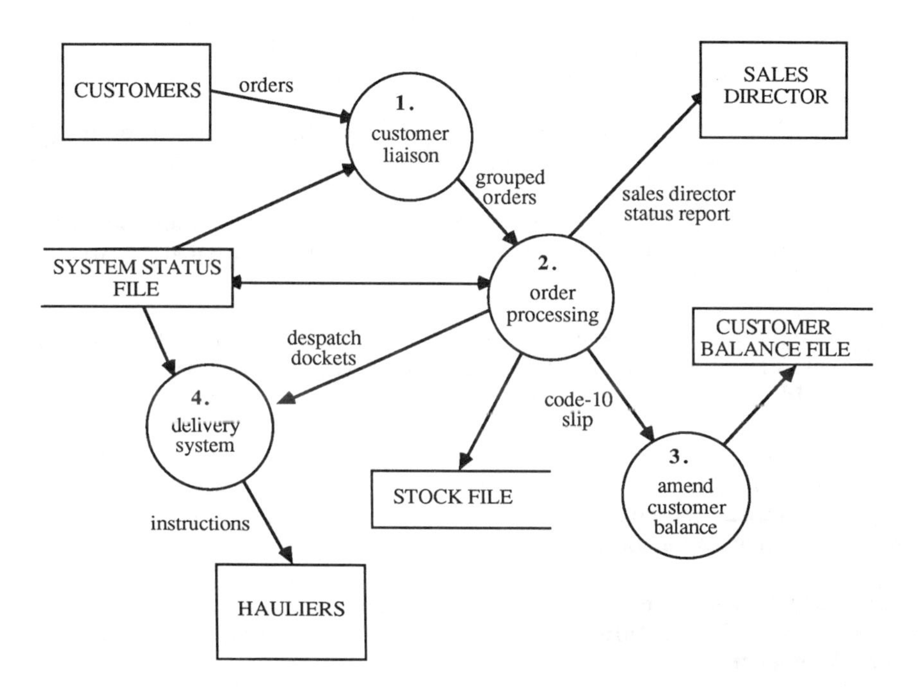

Figure 6.1: Physical Detail in a Data Flow Diagram

A possible case could be the system status file shown in figure 6.1. It is likely that a file with this type of name and data flows to and from different parts of the system probably contains a wide range of information most of which is totally unrelated. In cases of this kind, the historical groupings should be decomposed, thus exposing the distinct components.

For example, the *system-status-file* might have data dictionary entries such as:

system-status-file = { [free-vehicle-record, customer-record, outstanding-order-details] }

free-vehicle-record = registration-number + type + capacity

customer-record = customer-number + customer-name + customer-address

outstanding-order-record
= order-number + customer-number + { product-number + quantity }

Analysis of these entries (together with any supporting entity-relationship model) shows that there is no relationship between any of these data items and it could reasonably be assumed that the data has been placed into a single repository of pragmatic implementation reasons. This file can therefore be split into its logical constituent parts.

Political. Political considerations can often be detected through the use of an individual's name or general purpose processing. An example of political considerations in figure 6.1 is the process *customer liaison* which, upon analysis, may transpire to be something of a non-function. In this example, it may be justified to leave in process 1, although remaining would be necessary. Similarly, the *sales director's status report* has the characteristics of being purely political in that it refers to a specific person within the system.

The general strategy for removing physical considerations is to start with the top levels of a data flow diagram set, since this is typically where most problems occur.

In some cases where a process is inappropriately named, such as *delivery system*, which is too procedural, a review of a lower level data flow diagram will often reveal that simply the name is inappropriate and that the lower level diagram can be substituted for the process.

Figure 6.2 shows a revised version of the data flow diagram shown in figure 6.1. Notice the expansion of process 4.

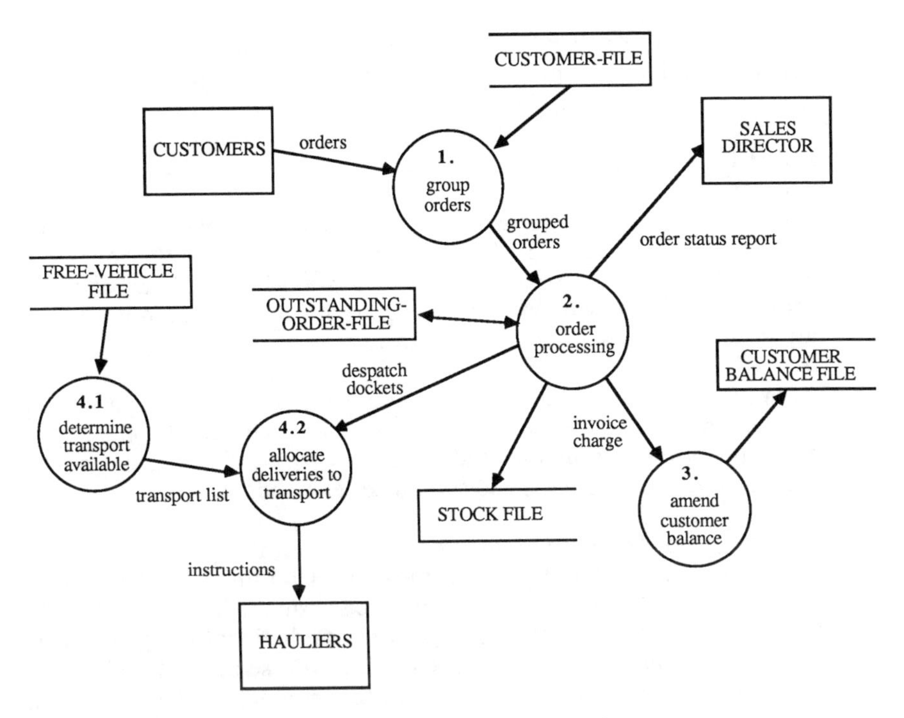

Figure 6.2: Logical Data Flow Diagram

6.3 Specification Validation

Before the transition from analysis to design can be completed, the analyst must be sure that he has accurately and completely specified the characteristics of the system that its future users require. It has already been said that analysis is not a process in which the analyst follows a straight path from a clean sheet through to a complete specification. The nature of analysis demands that the process is iterative, each iteration feeding on the knowledge gained to date and producing a more refined, accurate and complete view of a system. So whilst we shall talk about the validation of requirements specifications by users occurring prior to the start of the design process, such validation will be happening throughout the analysis phase and it is important that this is not forgotten.

6.3.1 The Purpose of Validation

The general purpose of validation, that of determining whether a specification meets its potential users' requirements, has already been identified. However, it is helpful to identify the particular issues with which the process of validation should be concerned. The main issues are as follows.

Acceptability. Acceptability is concerned with ensuring that all users involved in the specification of a system find that the statements, policies and models forming a specification are acceptable and do not conflict with their individual view. The presence of conflict does not necessarily imply that some aspect of a system is incorrect, but in all cases of conflict the analyst should regard this as a signal to undertake further analysis.

Accuracy. Accuracy relates to the precision to which a particular requirement or characteristic is documented and whether it can be open to interpretation. Ambiguity and vagueness will lead to problems later in the development process and should be removed.

Completeness. For any system, the degree to which its specification fully covers user requirements system must be determined. Checks at a syntactic level can be made by ensuring consistency between data flow diagrams, process specifications and the data dictionary, together with ensuring that data flows are balanced. However, semantic checks, such as whether a process has been omitted completely or a process specification is inaccurate, can only be determined through user validation.

Scheduling and Cost Control. The issues of scheduling and cost control are included here, not because they necessarily form part of a specification, but rather to trigger a check on the validity of any assumptions that were made at the start of the analysis. For example, if it is expected that design and implementation will take three times as long as analysis and analysis has overrun, the issue of how will this effect costings and timescales of the project must be addressed. Previous cost-benefit analyses may become invalid and require reappraisal.

6.3.2 Walkthroughs and Formal Reviews

Walkthroughs and formal reviews are two techniques which can be employed at any stage throughout the software development process. Their purpose is to examine project deliverables, such as specifications, designs or even computer programs, and determine at an early stage whether they contain errors. It should be noted that there are many variations of walkthroughs and formal reviews, but their underlying principles are the same as those described here.

Walkthroughs

The term walkthrough has been used to describe many activities but we shall use it to describe the case in which analysts present to users, aspects of their specifications, in an informal manner. Most usually, this is achieved by the analyst talking through a set of data flow diagrams and entity-relationship diagrams and discussing aspects of process and data specifications. Throughout the walkthrough, the user interacts with the analyst, correcting points of details and providing the analyst with further information where necessary. It is this interaction and relative informality which is the strength of a walkthrough.

When planning a walkthrough, the following points should be borne in mind.

- The majority of walkthroughs involve discussion of policy and procedure and are essentially of a technical nature; therefore participants in a walkthrough must be qualified to comment and have a sufficient understanding of the area.

- Walkthrough groups should be kept small, normally between 2 to 4 people, so as to give everybody an opportunity to comment.

- Walkthroughs should be kept to about 20 minutes in order to keep everybody's attention.

- Where representations such as data flow diagrams, structured English and entity-relationship models are used, the analyst should ensure that users understand the principles and conventions used, otherwise the user may become alienated and feel unable to comment on the specification.

- The walkthrough should aim to simply identify problem areas and not to discuss possible solutions which will always be a time-consuming process and usually based upon insufficient detail; solutions are better considered outside the meeting.

- A set of informal minutes should be produced, recording the main problem areas, together with some indication as to their relative importance in order that the analyst can address the most important difficulties first.

Formal Reviews

Formal reviews can serve many purposes, but their primary role is in establishing the attainment of project milestones. Within the context of a requirements specification, milestones include such events as the completion of a particular level of data flow diagram or the specification of the processes relating to one data flow diagram (see chapter 10 for a further discussion on milestones).

In the case of a formal review at which a complete specification is presented, the review represents the point at which the specification becomes the contract between those who have commissioned the development and the system developer.

Unlike a walkthrough, the formal review is largely a public exercise at which analysts attempt to demonstrate the competence with which they have created the requirements specification and that this specification meets the needs and desires of users. This is in contrast to the informal walkthrough in which the analysts' role is that of a learner who is both confirming and expanding his knowledge of system requirements.

A formal review has the following characteristics.

Size. The size of a formal review depends largely upon what material is under review, who will be effected by the proposed system and who else may have an interest in the project. Practically, at least 3 or 4 people should attend a review.

Composition. In addition to the analyst or team who has produced the review material, additional possible attendees at a review include:

- representatives from those who will be affected by the proposed system and who have contributed to the production of the material
- people who, whilst not directly affected by the proposed system, are related to those who will be, such as people under the same management
- specialists in reviewing such as system auditors or quality assurance staff
- outsiders who may provide unbiased views or have a breadth of experience beyond the scope of the immediate system, helping to see the problems and advantages of the proposed system, in perspective.

The Review Leader. Each review must have a leader who is not an analyst and is appointed by management. The role of the leader is to see that the review is properly conducted, ensuring that everybody has the opportunity to contribute and that all material for review is covered.

The checklist for a review leader must include:

- qualifications: do all those present, including the review leader, understand the specification and the purpose of the review

- is the specification or part of the specification ready for review, ensuring that all necessary supporting documentation and evidence is available for consultation

- have all aspects of the specification been reviewed

- to what extent is there agreement on the outcome of the review

- what action must be taken after the review.

The Secretary. Each formal review should be documented by a secretary or scribe; somebody who is capable of recording the salient points discussed. It is important that the secretary is not one of the active review participants, since in such cases they can be so busy distilling and recording comments that they have little or no opportunity to make their own contribution.

Further material on reviews and walkthroughs can be found in *Yourdon (1977)*.

6.4 The Objectives of Design

The transition from analysis to design marks a significant change of direction in the development of software and is summarised pictorially in figure 6.3. Prior to design, emphasis is placed upon the decomposition of a problem into a set of discrete, but logically connected statements of requirements and characteristics of a new system. This set of statements is embodied in a requirements specification and represents the agreed view of the functionality of a system, omitting physical detail and considerations.

Design however, marks the point at which the statement of requirements and characteristics is transformed to produce a physical system capable of supporting those requirements and exhibiting the desired characteristics. Design is thus a process of assembly in which groups of functions and data are brought together and whose requirements and characteristics are achieved through the definition of procedures and data structures respectively. Design does not concern itself with the actual production of software and data structures; this is the domain of the software implementor. The importance of design lies in the identification and evaluation of alternatives which satisfy a requirements specification.

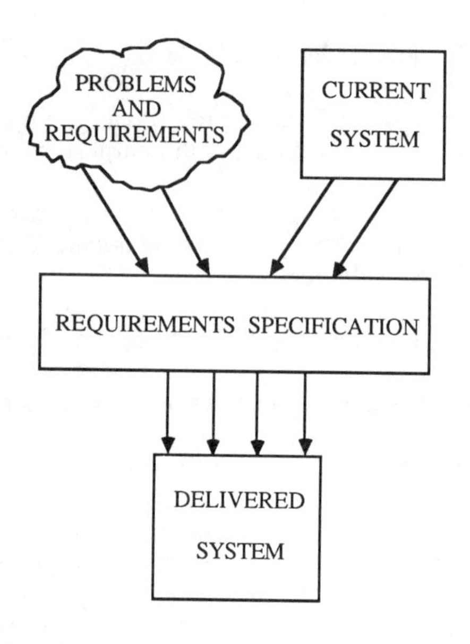

Figure 6.3: Overview of the System Development Process

The major objectives of system design can be summarised as follows.

Identify a System Architecture. Identifying the basic system architecture involves the definition of the shape of the proposed system. So far in the development process, the proposed system has only been defined in abstract terms relating requirements. System architecture, on the other hand, is a broad statement of how the requirements of the system will be achieved. A necessary task within this step will be the identification of available hardware, software tools and software packages and a specification of the alternative ways in which these could be combined to achieve an implementable system.

Identify Security and Safety Issues. In many systems, if they have not already been considered as major requirements, the issues of security and safety must be addressed. In terms of security, the system developer must consider the protection of the system and its data against fraudulent or unauthorised use. Safety, on the other hand, relates to the maintenance of copies of the system and its data for use in the event that they become corrupted, either through accident, such as a hardware failure or program error, or deliberate intent.

Identify Operational Requirements. Because of the nature of a specification, the task of automating a system must also include the addition of controls and procedures for running the new system. For example, the developer must define time schedules for when particular applications are to be run. Applications which perform complex updates on large volumes of data would probably when there is little other processing, so as not to disrupt other work. Such operational requirements will not normally be included in a specification since they relate to how a system will work, rather than what it is supposed to do.

Identify and Evaluate Alternatives. For any given system specification, there will always be a number of alternative implementation strategies. It is therefore necessary to identify the various possible alternatives and produce an appraisal of the costs and benefits of each alternative, since some of the alternatives may fulfil user requirements more completely or accurately than others, with a trade-off against price. On the basis of all the information produced to date, the best alternative can be selected for implementation.

Develop a System Design. Once the basic approach to system implementation has been identified, it is possible to develop the detailed design aspects, such as process design, database and file design and interface design. Processes will be documented in terms of program modules, data in terms of conceptual schemas and interfaces in terms of screen and report layouts, dialogues, coding systems and data validation procedures.

Review and Approve the System Design. Once the overall system design has been completed, together with the plan for its implementation, the development can be reviewed and approved, with modifications where necessary.

Chapters 7, 8 and 9 will look in detail at the process of developing a detailed system design, whilst the remaining sections of this chapter review system architecture, security, safety, operational design matters and alternative evaluation.

6.5 System Architecture

6.5.1 Boundary of Automation

So far, the process of software development has covered the stages of identifying user requirements and the specification of the desired system. The next step that must be undertaken is to define the boundary of automation, that is, how much of the system will be computerised. This is followed by the preparation of physical alternatives which will form the basis for establishing the eventual cost of the system. These alternatives are subsequently reviewed by the development steering committee and the best match between requirements and affordable solution is selected.

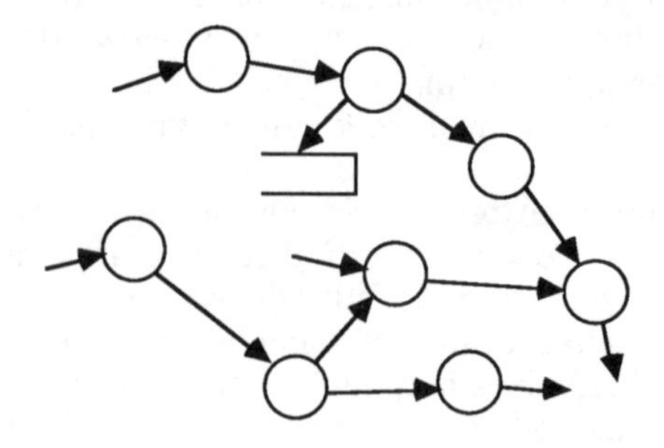

Figure 6.4: A Logical System Specification

Data flow diagrams may be used to provide a high level view of the system alternatives such as the diagram shown in figure 6.4. Using a data flow diagram as a basis, a system developer may devise various implementation alternatives, such as those shown in figures 6.5(a) and 6.5(b). The dotted lines represent the boundary of automation and everything included within this line is regarded as a system component to be computerised.

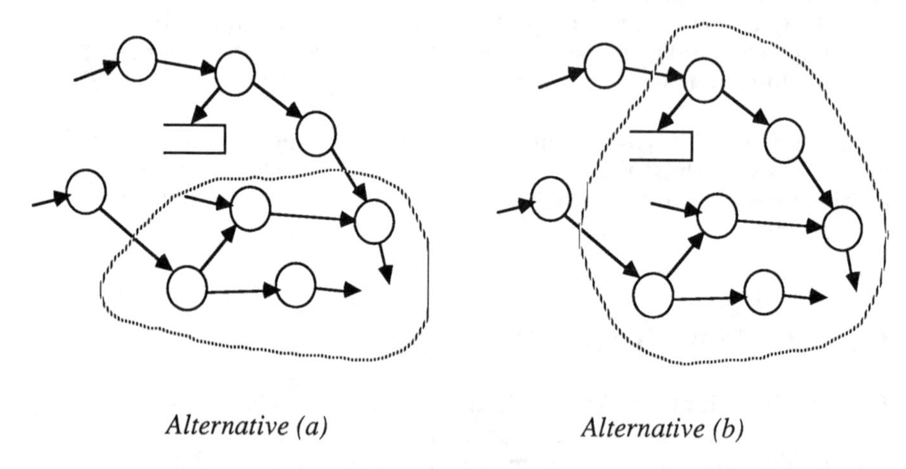

Alternative (a) *Alternative (b)*

Figure 6.5: Alternative Boundaries of Automation

Corresponding to these alternative boundaries of process automation will be alternative entity-relationship models.

6.5.2 Batch or On-line System

Once the boundary of automation has been agreed, the processing mode of a system must be identified. In particular, the system developer must decide whether the business with which the system is concerned requires immediate processing of activities or whether it is more appropriate for such activities to be grouped and processed in batches. These two systems of processing are referred to on-line systems and batch systems, respectively, and the following sections consider each type in turn.

Design Criteria for Batch Systems

In batch processing, transactions of the same type are collected and ordered into groups which are subsequently processed at some predetermined time. This time is defined by the developer of the system who takes into consideration the operational characteristics of the transactions and the user requirements.

Batch processing was the first processing mode to be used within computer systems, but nowadays it is hard to find pure, batch, data processing systems. However, there are many applications which are naturally batched, for example payroll, and therefore this mode of processing is often found in contemporary data processing environments all be it in a restricted format.

Batch operations need to be planned for exact times and for large applications which may involve many files and programs, careful scheduling of resources is of paramount importance. A systems analyst needs to consider the following points when designing a batch system.

- Define the way transactions will be captured and how the data will be converted to a computer readable format.

- Define the files and programs which must be loaded from the installation's library during a batch run and the sequence in which the programs must be executed.

- Determine procedures for backing up master and transaction files during a batch run. Backup must be effected at strategic points during the run so that computer resources will be used optimally and recovery, following a computer or software failure, will not involve unnecessarily long delays.

- Determine the type of personnel needed to operate the system and arrange for multiple shifts if necessary.

Design Criteria for On-line Systems

Unlike batch processing, on-line or real time processing does not require transactions to be batched before they are input to a system. However, it does require all files and programs that are needed to process the transactions to be constantly and directly accessible. In general, on-line systems may be considered as having a number of terminals connected via a network and communications controller to the central processor and the unit of work is known as a transaction. In a batch system all transactions are available for a computer run. By contrast, in an on-line system transactions are handled one at a time. The developer of such a system has less opportunity to optimise the system and thus faces greater design pressures in order to meet performance targets.

In designing an on-line system, the following factors must be taken into consideration.

Work Load. It is important to determine the work content of each transaction and the patterns of arrival of transactions both at the terminal and the computer with more detail than in batch systems. Each type of transaction has some work content associated with it. For example, a simple enquiry may require a single access to a file, whereas a complex customer order may require accessing many different files all of which must be available during transaction execution.

Variations in Work Load. An on-line system should process a transaction in a brief time after its arrival and when designing an on-line system care should be taken with regard to this processing time. Critical to this issue is the arrival pattern of transactions. In practice three key patterns of variation have been identified: seasonal, daily and random.

- Seasonal loading. If a system is designed to handle peak seasonal loading providing fast access time during peak transaction arrivals, there is going to be excess capacity during slack or trough periods. A batch system can cope with such peaks and troughs by careful scheduling, delaying work which is not critical, working extra shifts or even commissioning work to others. But an on-line system runs the risk of severe cost penalty in peak seasonal handling.

- Daily Pattern. In batch systems, a daily transaction pattern makes little difference (monthly, weekly, or even daily peak patterns can be handled in similar ways). However, on-line systems have to consider yet another peak period, this time the peak daily loading. It should be emphasised that daily load will seldom be constant.

- Random Arrival. Even within the daily peak periods fluctuation of arrival rate occurs. This arrival rate is random in its nature. It can occur at the terminal, the central computer or in other hardware components.

The problem from the developer's point of view is how much capacity to allocate so that the system can cope with the various arrival rates.

If the developer designs the system with a capacity which can cope with any possible load, the system will be expensive and perhaps under-utilised for long periods of time. Alternatively, by providing for the minimum loading, there is the danger of under capacity. The answer lies somewhere in between but the optimum solution requires careful investigation on the part of the developer.

Handling the Loading Problems. Loading problems arise from peaks (seasonal, hourly, or random). Lines of solution are similar and the developer can just consider the peak hour problem in isolation. The objective is to operate the system with a computer whose capacity is less than that required at the peak demand. Some solutions are outlined here.

- Planned overload. The developer provides insufficient capacity to meet the peak. This leads to queues but controlled delay may be preferable to wasted capacity. This must be treated with care and be acceptable by the user. In practice this can be dangerous as overloaded software systems tend to break down.

- Removing the peak. Peaks arise both from business activity and from the amount of on-line work included in a specification. Another solution to solving the peaks is by delaying selected transaction types or by delaying elements of transactions.

Other Considerations. A number of other factors that need to be considered are mentioned here. They are as follows.

- Response time. Every on-line system needs a response time target to be set before design is undertaken. Care must be taken in choosing the target because a tight target may lead to unnecessary cost of design and implementation, whereas too slow a response may cause user dissatisfaction.

- The currency of the data. Ensuring that the data made available to on-line systems is an important factor and system developers should ensure that the most up-to-date data is available.

- User contact. Because on-line systems are typically used by non-computer professionals, it is important that they are robust and reliable in terms of their software.

6.5.3 Distributed Systems

In addition to conventional batch and on-line systems, a new type of system implementation technique has evolved, the distributed system. Such systems have the following characteristics:

- they consist of two or more geographically dispersed computers on which applications programs can be run

- the computers are linked through telecommunication facilities

- the network of computers serves a single organisation.

The major factors that influence the design of distributed data processing systems are discussed in the following sections (*Reynolds, 1978*).

Centralised Versus Distributed Processing

The role of a central computer is difficult to define, since its presence and function within the system depends upon the kind of operations carried out by the organisation concerned. In most cases the host computer is a mainframe with facilities for storing large amounts of data, together with other peripherals such as printers and connections to terminals, similar to the example shown in figure 6.6. In its simplest form it may be a central machine acting as a message switch between remote computers. In other cases, the central computer may be used for storing corporate data files, performing complicated data processing operations that the smaller remote machines are unable to undertake or printing for voluminous reports.

Depending on whether or not a central node exists, the control of the system may be central or distributed. Control functions include:

- the handling of messages between the nodes

- the synchronisation of update operations

- the instructions for carrying out recovery procedures.

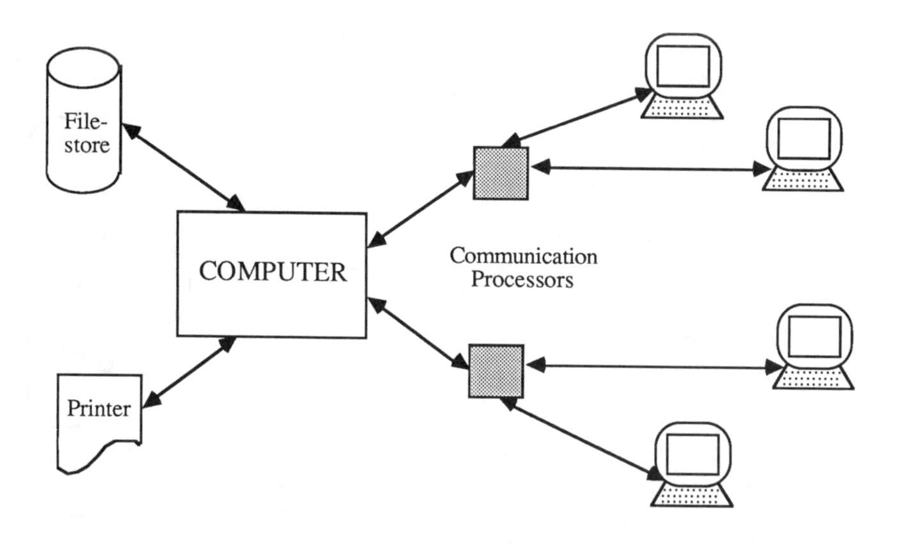

Figure 6.6: A Centralised Computer System

In a centralised configuration, control is handled by the central computer, whereas with the distributed approach, each node operates at an equal logical level and control is performed by the node that initiated the action.

Figure 6.7 shows a example distributed computer system. Notice that not all computers have all facilities; computer 2, for example, is simply a file server for other machines, whilst computer 3 organises printing.

Data Storage

The distribution of data can be achieved either by *partitioning* or *replication.*

A partitioned database results from dividing the total data into disjointed parts and assigning these parts to different nodes so that the data do not overlap. The allocation of a set of data to a particular node would depend on the processing requirements of the node under consideration, and usually, the data most frequently accessed are assigned to this node. For example, if a stock control program on computer 1 frequently accesses data pertinent to the site served by computer 1, obviously the relevant data will be stored on that computer. It would make very little processing and economic sense to store this data at another node since this would result in slower accessing and unnecessary communication costs.

Data can be partitioned in two ways (*Davenport, 1981*; *Gross et al, 1980*).

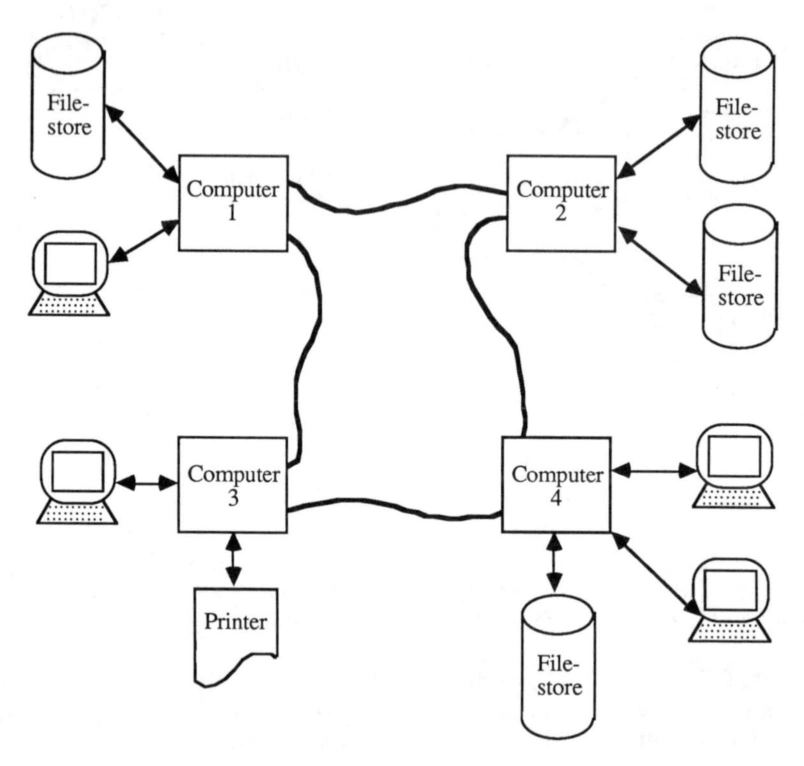

Figure 6.7: A Distributed Computer System

Data Partitioning By Value Or Occurrence. Using this method two or more locations may store the same record types, but only those record occurrences that are relevant to each location may be present at that location. For example, consider a company in the distribution trade which has one central office that deals with all the accounting and a number of different regional offices which deal with the the receipt of customer orders and the distribution of goods. Some of the data pertinent to this situation are *customer*, *product*, *order*, *invoice* and a logical structure for these is shown in figure 6.8 This structure is common to all regional offices but the record occurrences are not. That is, the *customer* records stored in one regional office are different to the *customer* records stored in another location.

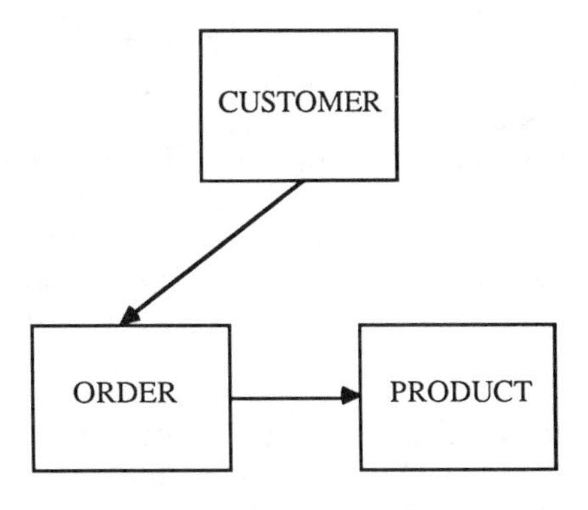

Figure 6.8: Data Partitioned by Value of Occurrence

Data Partitioning By Structure. In this method the record types are separated. The logical structure is divided into sub-structures between two or more nodes. The total structure is the sum total of all these sub-structures. This concept is shown in figure 6.9 which shows *invoice* being one of the data types held at the head office but not at any of the regional offices. If details about any of the invoices is required at a regional office, the corresponding information must be accessed by referencing the files at the head office.

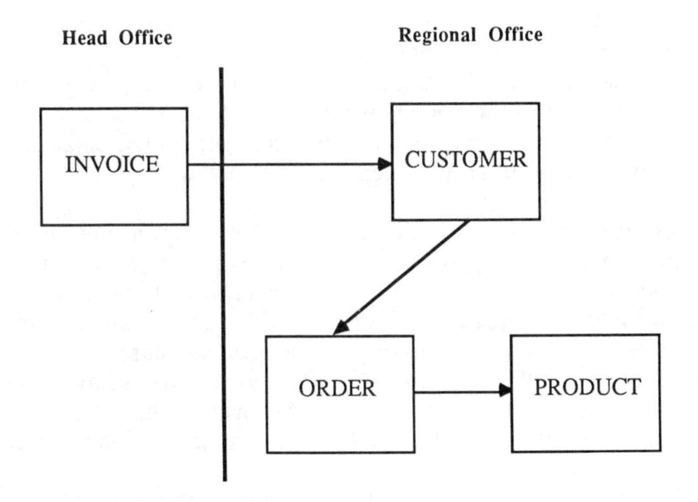

Figure 6.9: Data Partitioned by Structure

A replicated database results from the existence of the same sets of data in more than one location. A replicated database may be fully replicated, where every node has a copy of the total data, or partially replicated. The advantage of replicated data over partitioned is the improved performance with regard to accessing the data (since the data will normally be present locally) and this, in turn, results in reduced communication costs. However, updating of data may create problems, since updates to a set of data have to be implemented on all copies at every location and also, overall storage costs are higher.

There are no easy rules which govern the choice of data distribution. Small amounts of data that require very little updating are normally better distributed in a replicated manner. For small amounts with high updating rate, partitioning may be chosen. The factors that would have to be considered are the amount of data, the frequency of updating and the frequency of accessing local data (*Champine, 1980*).

Transaction Processing

When a transaction is processed, it is necessary to access a set of data on which application programs carry out a set of operations. In a distributed system there are cases where the required data will not be stored locally, resulting in the need to access data from a different node. There are three ways of handling the processing of transactions in a distributed environment.

Transaction Switching. With transaction switching, the transaction itself is moved to the location of the data. Here an applications program accesses all the required data, processes the transaction and returns the results back to the node that initiated the transaction. No movement of data occurs. A partitioned or partially replicated database is the most appropriate data arrangement for this method.

Transaction Splitting. With transaction splitting, a transaction is divided into a number of parts, each of which is processed within an individual node accessing data (if necessary) local to that node. When a part of a transaction is processed the intermediate results are sent on to the next node for further processing. The process carries on by forming a chain of partial transactions. When all the parts have been processed the final result is sent back to the originating node. A partitioned database is the most suited data organisation for this method.

Local Transaction Processing. A transaction is processed at the node where it originates. If it requires data from a remote location, this data is transmitted to the node executing the transaction. This method is applicable to both partitioned and partially replicated databases.

6.6 System Security

The problems of damage to computers intentionally or otherwise has been well publicised in recent years, with fraud and abuse of computer systems costing organisations a high price. Furthermore, failure to adequately secure computer systems can effect the rights of individuals and it is this concern in particular which has caused many countries to adopt national legislation to protect and regulate the data held within a computer system. In the UK, the Data Protection Act establishes eight principles for the protection of data which all computer installations must abide by. As most of these principles impinge upon issues of system security, they are listed in figure 6.10.

Data Protection Act Principles

1. The information to be contained in personal data shall be obtained, and personal data shall be processed, fairly and lawfully.
2. Personal data shall be held only for one or more specified and lawful purposes.
3. Personal data held for any purpose shall not be used or disclosed in any manner incompatible with that purpose.
4. Personal data held for any purpose shall be adequate, relevant and not excessive in relation to that purpose.
5. Personal data shall be accurate and, where necessary, kept upto date.
6. Personal data held for any purpose shall not be kept longer than is necessary for that purpose.
7. An individual shall be entitled to be informed by any data user whether they hold personal data of which that individual is the subject and, where appropriate, to have such data corrected or erased.
8. Appropriate security measures shall be taken against unauthorised access to, or alteration, disclosure or destruction of, personal data and against accidental loss or desctruction of personal data.

Figure 6.10: The Eight Principles of the UK Data Protection Act

Compromises to system security can result in three possibilities:

- loss of data availability, with consequential delay in the operation of an organisation and delays in the servicing of its information needs
- loss of data integrity, causing inconsistent and erroneous data
- loss of data confidentiality.

Consequently any system development must consider precautions against these losses. It should be noted that no system can ever be guaranteed to be totally secure, however, there are several measures that a system developer can take in order to minimise the risk of loss of data.

Over a period of time, the more significant threat to an organisation may be from the frequent occurrence of breaches which individually cause relatively minor losses rather than from the isolated large breach.

During system development, all possible breaches of security must be examined and a strategy developed to prevent or minimise any losses which may result from a breach of security. No general rules can be set out to govern the selection of security procedures. The controlling factor is cost and this cost needs to be balanced against the time wasted in case of a breakdown and the ensuing disruption to computer services. The following measures can be considered.

6.6.1 Physical Security

Physical security is concerned with the security measures taken at the computer centre itself. Some natural disasters are difficult to protect against but with many the risk of damage can be reduced by taking precautions against fire, flooding etc.

Physical security applies not only to protecting against damage but also to protection against theft and this is particularly relevant for small business systems. The main security measure against this happening is through the use of physical access control devices, which may vary from keeping computers or data storage media (disks and tapes) in a locked room, through to more sophisticated computer-controlled door locks which are opened by users entering a numeric code on a keypad located beside the door.

6.6.2 Access Security

The theft of data, via copying or reading is just as a much a cause for concern as its physical theft. Protection can be at two levels; preventing the thief gaining access to the system and protecting the data itself.

A key or badge may be used to connect a terminal to its host computer. Such a key can either be physical, requiring its insertion into a lock or conceptual, such as the entry of a user number and password, a system employed on most mini and mainframe computers.

Perhaps the most familiar safety device to most readers will be cash machine cards and their associated PIN numbers. With this system, machine users must insert a pre-programmed card into a reader and enter a secret number associated with the card. Only when the two match, will the user be permitted to perform transactions.

6.6.3 Transmission Security

A final category of security that should be considered, especially in a distributed system, is that of data passing along communication channels. The particular problem here is to prevent the interception of data passing along the channel, both to protect its privacy and to ensure that data is not altered, deleted or added to.

Several techniques can be employed to secure transmitted data.

Data Encoding. Data encoding can be used to protect access to data, prior to its transmission along communication channels. Subsequent decoding will be necessary upon receipt of data.

Message Numbering. In order to prevent the addition or deletion of messages passing along a communication channel, a system of message numbering can be employed, where each message sent between two nodes in a communications network is assigned a consecutive message number. In the event of a message being added or deleted, the numbering sequence will be broken and a violation detected.

6.7 System Safety

In any system concerned with the recording and processing of data, comprehensive file backup is needed for file reconstruction following the corruption of data. It is the responsibility of the developer to ensure that the ensuing disruption of services, following a system breakdown, is minimised.

System breakdowns are unavoidable and there is a need for a mechanism which will enable the restarting of the system, ideally at the point of breakdown. This mechanism is provided by the presence of a set of backup and recovery procedures which automatically recover production programs, system software and data files after a system failure.

The simplest form of backup and recovery procedure is the generation technique. This involves the saving of the files relating to some previous transaction periods to be retained. This gives scope for reprocessing all the transactions starting at any point within the saved period.

A better method however, is the use of checkpointing, which has the advantage of preventing the whole computer run having to be reprocessed. It involves the dumping of files and program images at various stages during an update, together with a log of the data updates applied by each running program. Figure 6.11 shows a summary of a checkpointing system.

The system works through three main operations. The first operation, is a periodic dump of the system's filestore to a tape. A message indicating this fact is recorded on a *journal tape*.

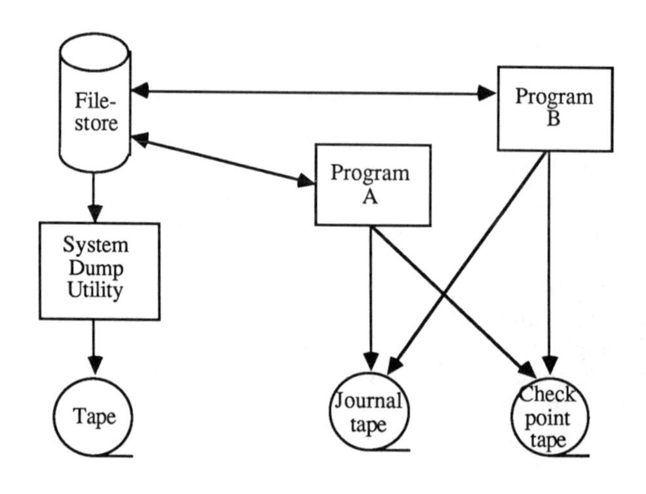

Journal tape

System dump	Record 23 updated	New record 24	Record 14 deleted	Checkpoint 5 taken	Record 19 updated	New record 34

Checkpoint tape

Start checkpoint 4	Image of program A	Image of program B	Start checkpoint 5	Image of program A	Image of program B

Figure 6.11: A Checkpointing System

The second operation, which again occurs at periodic intervals, although not necessarily the same, is a dump of the complete state of the computer system's memory, i.e. details about the running programs. These details are written to a *checkpoint tape* and a record of the event is also made on the *journal tape*. An inspection of the checkpoint tape in figure 6.11 shows that there have been two checkpoints, numbered 4 and 5.

The third operation occurs whenever a running program makes an update to the system's filestore. The program making the update records details of the update on the *journal tape*. Again inspection of the journal tape in figure 6.11 reveals updates to records 23 and 19, insertion of records 24 and 34 and the deletion of record 19.

In the event of a system failure, a *restart* operation occurs. This operation begins with the system searching the journal tape for the last system dump. Once found, the appropriate system dump tape is used to overwrite the system's filestore and so restore the filestore to a state prior to the system crash.

Next, the system must update the filestore by duplicating the operations performed by the running programs, upto the last checkpoint. Thus in figure 6.11, the system would have to update record 23, insert a new record 24 and delete record 14. Finally, the system would find details of the checkpoint on the checkpoint tape and restore, to its own memory, details of the running programs. In figure 6.11, this would be checkpoint 5. The restored programs can then be set running again from the instruction immediately after the last instruction they performed before the checkpoint was taken.

In many large systems, such checkpointing and restart facilities are automatically provided as part of the operating system and database management systems; however, system developers should ensure that they do exist and take their own appropriate measures if necessary.

6.8 Operational Design

In any system, a degree of auditing and control must be imposed which ensures that the system runs in the intended manner and performs the required functions at the correct time.

The types of internal control that may be found in an information system fall into two categories: operational controls and administrative controls.

6.8.1 Operational Controls

Operational controls are needed to ensure the efficient running of a system and include work scheduling, input, processing and output.

Work Scheduling. The sequence of work involved in the day to day running of the data processing system should be carefully defined to ensure efficient processing.

Input Controls. These controls are established to ensure that only authorised data enters the system and may be applicable to operations at the user or the computer department.

Processing Controls. Many checks may be carried out by a computer during operation without human intervention and processing controls fall into this category. Typical controls applied to data are those of batch totals, check digit verification, field interrelationship tests, presence tests, sequence tests and range checks.

During an update run controls may be established in the updating program which check the accuracy of the processing. Such controls include: trailer records; tape cycling; trailer records and dumping and logging.

Output Controls. These include manual verification of output and exception reports.

6.8.2 Administrative Controls

Administrative controls are designed to ensure that the data processing department operates efficiently and involves division of responsibilities (e.g. development, operations sections etc.) and control over the daily working of each division. With particular reference to operations controls include: the use of an operating log; printout of all operator interventions; the use of a manual outlining all operating instructions.

6.9 Evaluation of Alternatives

6.9.1 Packages

Inevitably, it will always be possible to transform a specification into several different designs and implementations. For example, the preceding sections have outlined the design alternatives of batch versus on-line, distributed versus centralised. However, there is perhaps a more fundamental issue that any system developer must ask, namely, should the proposed system be individually produced or should the purchase of an off-the-shelf, application package be considered.

Central to such a decision is an appreciation of the purpose of the proposed system and whether a pre-written package is available. In principle, this is not as unlikely as it may seem, since many administrative information systems will require similar functionality, usually consisting of procedures for:

- information storage and update
- information listing based upon simple selection criteria, such as matching values
- elementary calculations, such as averaging and summing
- querying individual items if information.

In addition, specific packages will contain procedures for performing particular tasks. For example, a payroll system will maintain payscale tables, staff details, calculate pay, deduct tax and produce pay cheques.

The obvious advantage of using an application package is the cost saving, since the purchase or lease price of a package, will normally be less than the full development cost. Other advantages include:

- quicker implementation, since a package usually already exists
- greater efficiency, since with mature packages, considerably greater effort will have been invested in their refinement and improving efficiency of operation
- maintenance can usually be obtained on a contractual basis for automatic updates as better versions are released or as legal, functional or environmental factors require changes to the system
- transfer to new hardware may be easier, since many packages are often implemented on more than one machine.

However, as well advantages, there are tradeoffs when using packages and these can be summarised as follows.

Generality of Package. The disadvantage of most packages is their generality in terms of a user's precise set of specifications. Most organisations tend to assume that their requirements are fixed and often reject the package approach because these cannot be exactly met. Whilst this may sometimes genuinely be the case, it is worth analysts and users carefully examining the requirements set to ensure that the *difficult* requirements, that cannot be accommodated by a package, are not there simply for historical or administratively convenient reasons. The benefits of a package approach may well outweigh these considerations and justify some change in requirements. In addition, many package vendors recognise this disadvantage and provide users with the ability to insert extra code or call user-written routines.

Interfacing. Another problem that can arise with packages is their inability to interface with other parts of a system, whether those parts are built in-house or are other packages. Obviously, in-house systems could be adapted to interface correctly, but this may not be so easy with other, externally-produced packages.

Performance. Whilst many packages have been refined and tuned to give good performance, the generality of packages both in terms of functionality and their ability to run on several different machines and architectures, may impose a performance overhead. For example, consider a word processor which recognises terminal function keys. Because the signals generated by function keys tend to vary between terminals, most word processors, rather than directly recognising a function key, will have a look-up table, against which they match a received code and hence identify the function key pressed. There is, therefore, an overhead arising from the word processor's ability to run on different terminals.

6.9.2 Implementation Technology

Assuming that a developer wishes to proceed with a customised system, as opposed to a package solution, this may be the stage at which to consider the selection of implementation technology. By this we mean what system packages and language might be used and how to produce the program code.

System Packages

System packages are similar to application packages, but are generally hidden from the eventual users of a system. Instead they are used by implementors to perform well-defined, repetitive tasks. Perhaps the most widely known system packages are database management systems, which automatically store, format and retrieve data through a well-defined interface, removing the burden of detailed storage and data organisational issues from implementors.

Other system packages, traditionally supplied with operating systems are filing systems, file sort and merge utilities and routines to obtain operating system data such as a list of the users logged onto the system, the current date and time etc.

More recently, many computer manufacturers and software houses have begun to supply management systems for screen handling, transaction processing and data access in multi-user systems. All these system packages are aimed at reducing the burden of the most tedious programming work and thus allowing the programmer to concentrate on broader issues such as the overall functionality of the system, performance and user interface design.

Code Production and Language

Inspite of application and system packages, there is always inevitably some part of a system that must be coded and the developer must therefore make a decision about how the code is to be produced and in what language.

With regards to code production, there are three alternatives.

Code Generation. An increasingly popular method of overcoming the cost and time of program coding is to use a code or program generator. These generators accept a variety of inputs, including entity-relationship diagrams, data flow diagrams, descriptive text conforming to a pre-defined syntax. The user of a generator must also indicate the type of program required. When activated, the generator will then produce a skeleton program, usually in a language such as COBOL, which the developer can edit, compile and run in the usual way.

Code Reuse. Over the years, many installations have developed large libraries of program code which have well-defined interfaces and which can be used to implement standard functions within application systems. For example, a banking system may well develop standard routines for currency conversion using different policies for commission deduction, rounding etc.

Direct Coding. If no other alternative exists, the last resort is to code a system directly.

Selecting a programming language for direct coding is often not such as a difficult decision as one might expect, given the wide variety of languages available. This is because most developers will have already be constrained to particular machines, supporting only a limited range of languages. Other factors will include the availability of languages to interface with database or screen management systems or program generators. It is now well recognised that five generations of language exist, and these are summarised in figure 6.12.

Each generation has it own particular advantage and disadvantage and these may be summarised as follows.

First and Second Generation Languages. These languages relate closely to the architecture of a computer and thus provide significant advantages in speed of execution and overall performance compared to other languages. Their main disadvantage is the inability to represent high- level constructs and programs written in such languages are generally difficult to read. The other disadvantage is that programs written in these languages are not portable between different ranges of computers.

First and Second Generation	Assembly language System/360
Third Generation	COBOL FORTRAN Pascal Ada BASIC
Fourth Generation	SQL Oracle
Fifth Generation	Prolog Lisp

Figure 6.12: The Five Generations of Programming Language

Third Generation Languages. These languages are still the most widely used languages and consist of sets of procedural instructions which are obeyed in sequence by the computer. Special constructs are provided for repetition of code, conditional processing and jumping. Languages such as COBOL are widely available and because considerable effort has been made to standardise implementations of the language across different computers, programs written in these languages are generally portable across computer ranges.

Fourth Generation Languages. These languages can be regarded as a cross between application and system packages. Typically such languages are general purpose query languages which operate on a user-defined database. Examples include NOMAD (*McCracken, 1980*) and ORACLE.

Fifth Generation Languages. These are the newest group of languages and represent a significant change in the style of programming. Traditionally, programming languages have always placed emphasis on how a computer is to perform a task. Fifth generation languages, like Prolog (*Clocksin & Mellish, 1984*) and Lisp (*Winston & Horn, 1984*), by contrast, concentrate on what is to be performed, relying upon underlying operational semantics to determine how it will happen. The effect is to greatly reduce the complexity of programming and provide a more natural approach. However, such languages, whilst gaining wider acceptance, still only have a small user base.

6.10 Summary

The transition from analysis to design, which marks a significant turning point in the development of software, involves several processes.

Firstly, the analyst must ensure that a requirements specification portrays only the logical view of the required system and does not attempt to preempt design decisions which relate to how the required system might be achieved.

Secondly, because a requirements specification will form the basis of a contractual obligation for system developers, the specification must be validated by its potential users and those who have made input to the analysis phase. This validation will initially take the form of informal walkthroughs, culminating in a formal review.

Once these tasks have been completed, the specification can be signed-off and passed onto the system developers. At this stage, the basic characteristics of the system must be identified, such as batch versus on-line, centralised versus distributed. Alternative implementation strategies must next be identified and once evaluated, a suitable policy adopted.

Finally, the detailed design of processes, data and interfaces can commence and the next three chapters will examine these processes.

Chapter 7

Process Design

The aim of process design is to specify a number of program modules and their interrelationships which, when implemented, will exhibit the characteristics identified in a requirements specification document. However, this is not sufficient, as the assembled modules must also be constructed in such a way that they are clear, readable and maintainable.

Two distinct approaches to process design exist: *process-driven*, in which modules are developed from a processing perspective and *data-driven*, in which modules are developed from a data perspective. This chapter examines the aims and techniques of these two approaches.

7.1 Aims and Objectives

Process design is about constructing a description of a set of program modules which when interacting in a computer system, will perform a required application. For example, figure 7.1 shows a module hierarchy for the production of hotel guest invoices.

In this section, the basic concepts of module description and good design are examined, together with an examination of why modules are structured as hierarchies.

7.1.1 Basic Concepts

The basic technique for representing modules is through the use of a *structure chart*, which is a graphical tool for representing the structure and relationships within a set of modules. Structure charts, therefore consist of two basic buildings blocks: *modules* and *connections*.

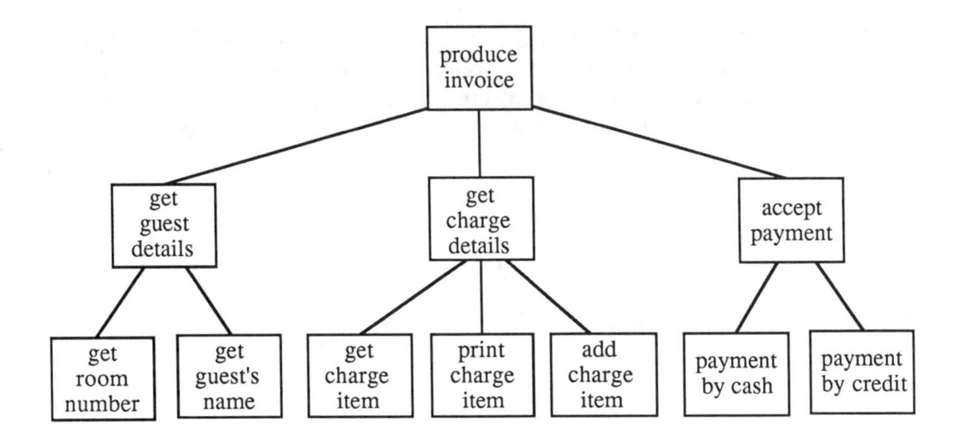

Figure 7.1: A Sample Structure Chart

Modules. A module is a bounded set of executable statements represented by a rectangular box containing the name of the module. For example, *get room number* and *get guest's name* are both modules belonging to the *get guest details* module shown in the structure chart in figure 7.1.

Connections. Connections between modules are represented by lines connecting two modules. Such connections occur where one module references (or calls, in programming terminology) another module. Thus the connections between *get charge details* and the three subordinate modules, *get charge item*, *print charge item* and *add charge item*, indicate that during the execution of *get charge details*, these three subordinate modules will be referenced.

In addition to the basic structure charting principles, additional notation can be added to refine the process description. By convention, the two process design approaches of process-driven and data-driven adopt different notation and these are briefly described in the following section.

Process-Driven Notation

Process derived structure charts, by convention, can show some of the following additional notation.

Parameters. Connections can be elaborated by the addition of parameters. These are shown by the presence of arrows pointing into or out of a module. Arrows pointing down into a module indicate parameters which are used to pass information to a module such, as a command code or data value, whilst arrows pointing up out of a module indicate parameters which are used to pass return information to the calling module, such as a status code or computed result. Figure 7.2 shows an example of parameters being passed to and from the module *get guest details*.

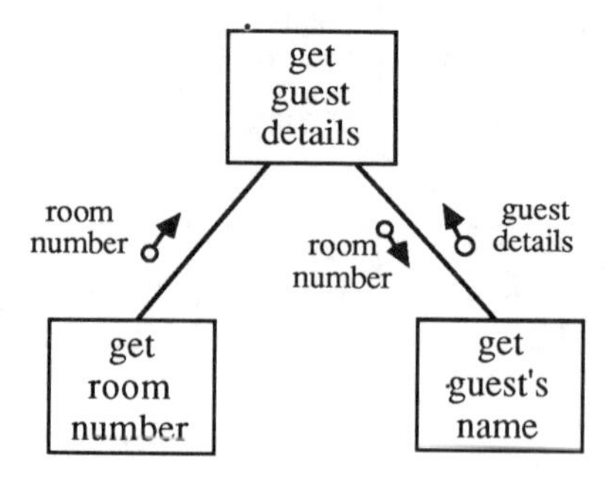

Figure 7.2: A Structure Chart with Parameters

A further useful distinction is the differentiation between parameters passing data for processing and parameters passing data for control purposes. Data parameters show ordinary data that is to be manipulated by a module and are shown as arrows with circles. In contrast, control parameters show data which will cause a module to behave differently depending upon the value of the data, such as a command code or menu option. Control parameters are shown as arrows with a dot on the end of their tail. An example is shown in figure 7.3.

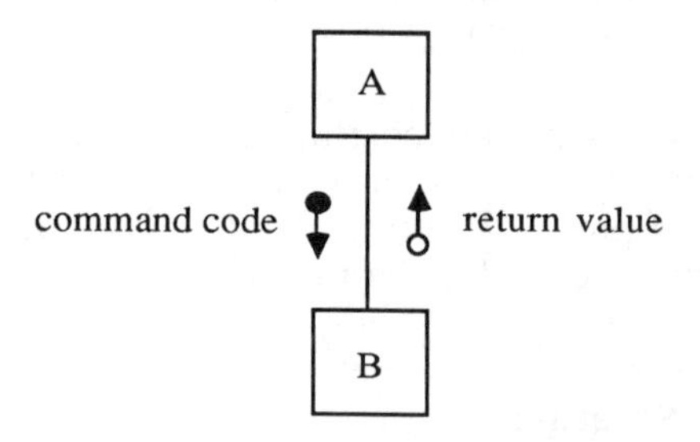

Figure 7.3: Control and Data Parameters

Procedural Detail. Additional notation can also be placed on a structure chart to document certain aspects of procedure. In chapter 4, it was noted that all procedures can be represented in terms of sequences, iterations and selections. Consequently, procedural annotation of structure charts falls into these three categories.

Sequences are represented simply by the order in which modules appear, thus in figure 7.1, *get room number* will be referenced before *get guest's name.*

Iteration is represented by the addition of a circular arrow enclosing the sequence of modules which are repeated. For example, in figure 7.4, the module sequence *get charge item, print charge item* and *add charge item*, will be repeated.

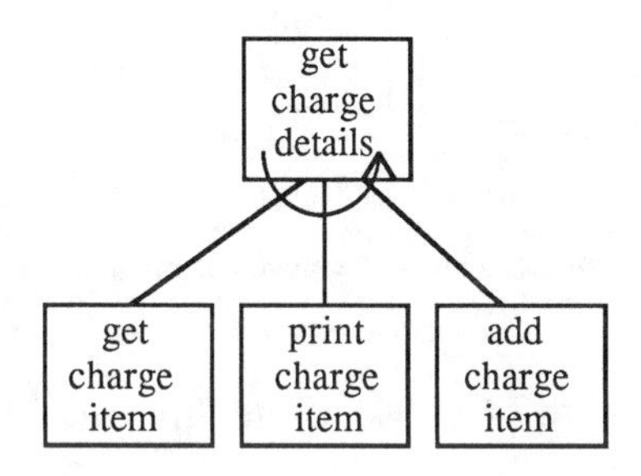

Figure 7.4: Repetition in a Structure Chart

Selections are indicated through the use of a diamond as shown in figure 7.5. The figure shows that *accept payment* consists either of a reference to *payment by cash* or a reference to *payment by credit.*

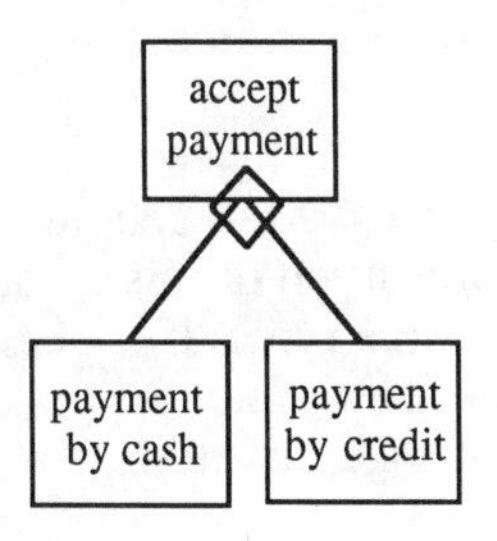

Figure 7.5: Selections in a Structure Chart

Figure 7.6 shows a revised version of figure 7.1 which incorporates all the additional structure chart annotation.

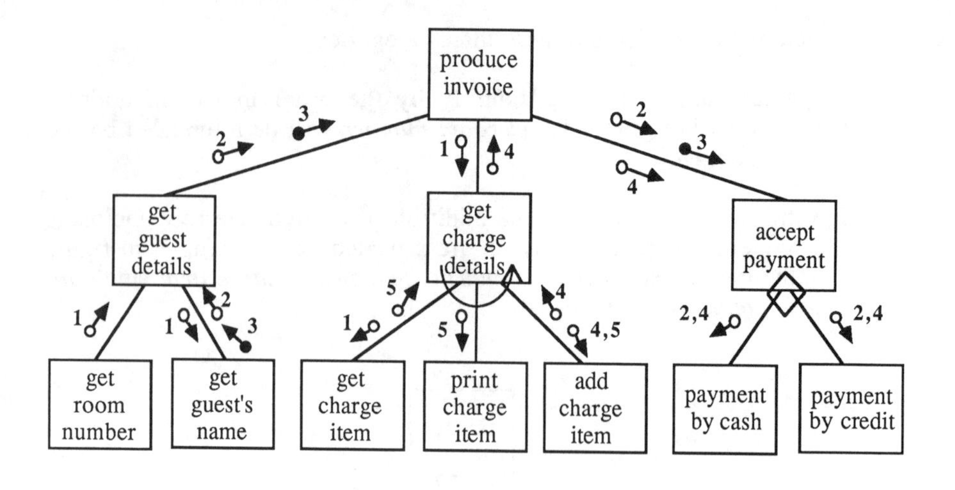

Figure 7.6: A Complete Structure Chart

Data-Driven Notation

The additional notation used for structure charts developed using a data-driven approach is the same as that used to describe task structures, previously described in section 4.2.

Figure 7.7 shows an elaborated version of figure 7.1 using the data-driven notation. Recall that asterisks indicate repetition and circles represent alternatives.

Notice that an additional process has been added: *process charges*. This is necessary because it is a convention of this notation for connected modules at the same level to be of the same type. If *process charges* had been omitted, *get charge details* would not be the same type as *get guest details* or *accept payment* and would thus infringe this rule.

7.1.2 Hierarchical Structuring

In previous chapters, the principle of *decomposition* was described as a mechanism by which analysts can handle and understand the complexity and

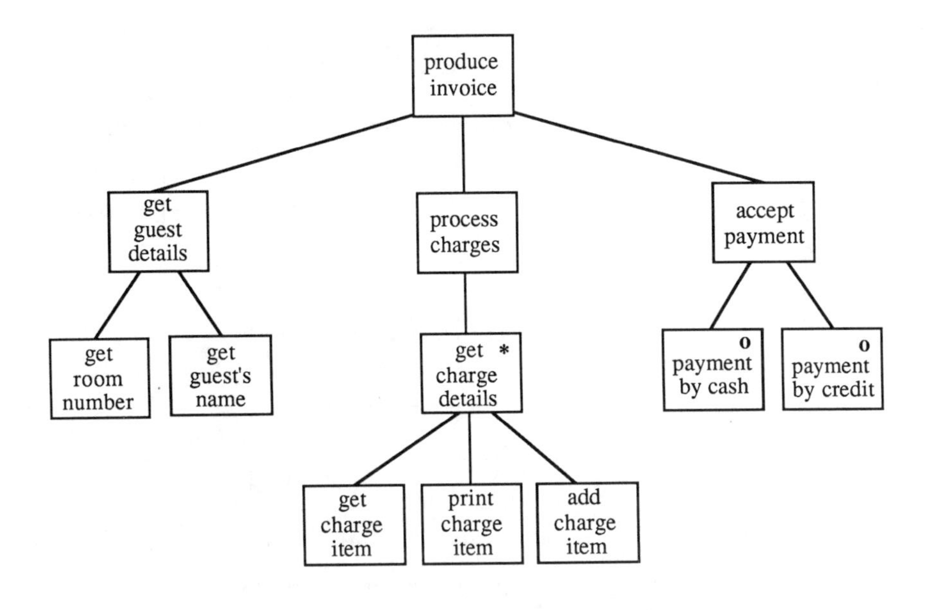

Figure 7.7: A Complete Structure Chart

scale of large systems. Similarly, designers also need techniques by which they can handle such systems and as with analysis, the principle of decomposition can be applied.

The application of the decomposition principle can easily be seen through the analogy of a business, in which different members of the business perform different functions. Figure 7.8 shows a sample business structure. The figure has several characteristics worthy of note.

- The business has a hierarchical structure. The higher levels represent members of the business who have a broad, general view of the business, whilst the lower levels represent members of the organisation who have a more limited view, but with a greater depth of knowledge.

- The higher levels of the business are generally concerned with control and policy of the activities conducted by the business, whilst the lower levels are concerned with the implementation and discharging of that policy.

In the same way that hierarchical structures permit the efficient organisation of a business or organisation, so they may be applied to the structuring of software.

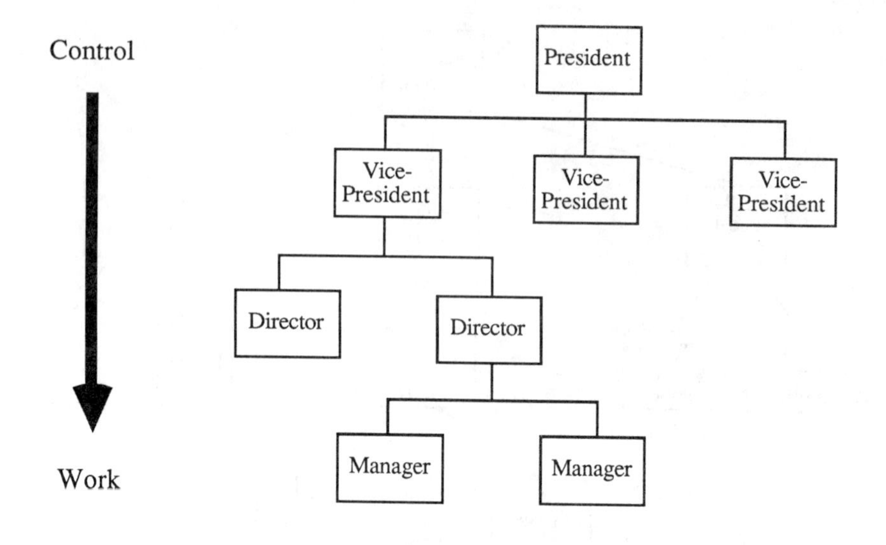

Figure 7.8: A Typical Business Hierarchy

The decomposition of a software system from high-level control modules down into low-level modules performing a series of data manipulations is an ideal and long-established approach to the handling of complex systems.

Such an approach has the following advantages:

- it permits easier design by allowing designers to decompose a system into manageable units
- it permits easier design by allowing designers to concentrate on the basic control mechanisms before developing the activity detail
- better project management and control can be achieved since once the basic hierarchical structure has been specified, designers can work relatively independently of each other
- testing is easier since individual modules can be tested in isolation of each other before the complete system is ready for assembly
- easier maintenance can result, since minor changes in specification can be implemented by the alteration of a small number of modules.

These advantages, however, only accrue when a system's module hierarchy is well structured. For example, if a particular system function is achieved through every module in a system performing some small task, the advantage of easier maintenance certainly would not apply since if the characteristics of the function was changed, potentially every system module would have to be changed.

The question which must be posed is how do we achieve the best decomposition. For example, figure 7.9 shows three possible decompositions of a particular system in which five basic activities must be performed. Each possibility exhibits different control characteristics- the question is which is best?

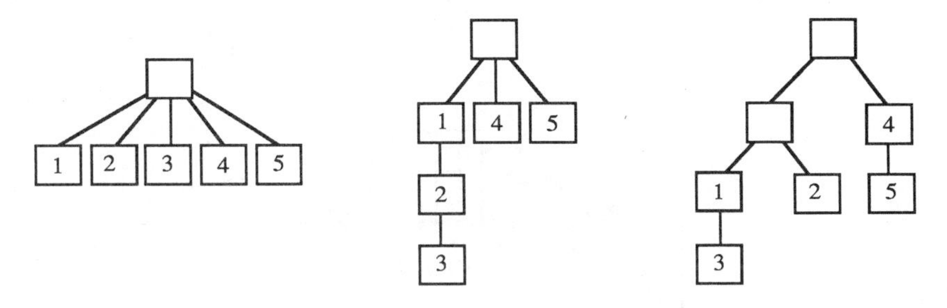

Figure 7.9: Alternative Process Structures

The key to unlocking the solution to this question relates to the issue of *module independence*, since the advantages of a hierarchical system only apply when each software module has a high degree of independence. The aim of process design is to therefore develop a set of module which have a high degree of independence.

7.2 Module Independence

Module independence is formalised in the concepts of *coupling* and *cohesion*. Coupling relates to the number of connections between modules, whilst cohesion relates to the functionality of a given module. These concepts are now further examined.

7.2.1 Coupling

Coupling, in its widest sense, is concerned with the number of connections that exist between two entities and hence the degree of independence between them. In the context of software design, the term coupling is used as a measure of the number of data links between modules, where a link may be defined as one reference by a module to data outside that module and which is necessary in order to perform its function.

For example, figure 7.10 shows a simple program hierarchy chart in which each connecting line represents one item of data. From the figure, it can be seen that module C has the most connections with module A, whilst module B has the least number of connections. From this, it can be concluded that module C, with the greatest number of connections to module A, will probably be the most dependent on module A.

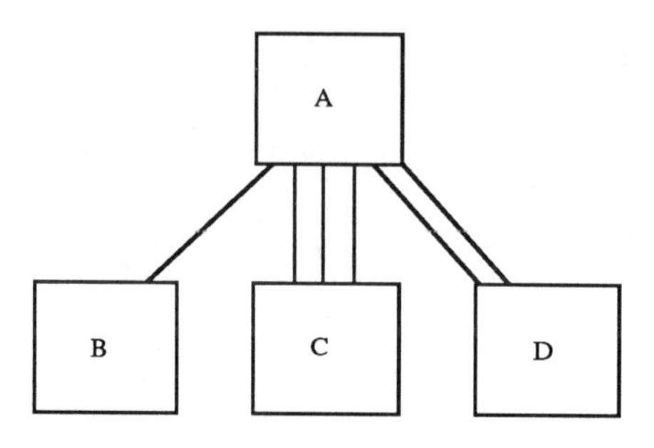

Figure 7.10: Degrees of Coupling

However, in determining the degree of coupling between two modules, it is not sufficient simply to look at the number of data connections- other factors such as the complexity and type of connection must also be considered. These factors which influence coupling are as follows.

Complexity of Interface. The complexity of an interface has a considerable bearing upon the degree of coupling between two modules. For example, compare two sets of modules, one of which has 2 parameters which are passed between modules and the second having 22 parameters. The former pair of modules are likely be less tightly coupled than the latter. In general, this will normally be the case, however, a simple parameter count can sometime be misleading. In some languages, such as COBOL or Pascal, parameters can be grouped together into records and passed as a single, physical parameter, when in fact that they represent several, logically independent parameters. Figure 7.11 shows an example of a complex parameter definition in COBOL. Whilst the first three fields may have some logical grouping, the four field, *op-code* has a logically separate role (it acts as a controlling parameter), as do the remaining parts of the parameter.

```
WORKING-STORAGE SECTION.
1     complex-parameter.
 2    sales-value           PIC 9(6)V99.
 2    saleman-count         PIC 999.
 2    average-sales         PIC 9(6)V99.
 2    op-code               PIC X.
 88   calculate-average     VALUE "C".
 88   print-details         VALUE "P".
 2    system-date           PIC 99/99/99.
 2    page-heading          PIC X(80).

                     :

CALL "sub-prog"
      USING complex-parameter
```

Figure 7.11: A Sample COBOL Definition

Parameter Type. The type of data passed as a parameter between two modules also plays a significant role in the measurement of coupling. Parameters can always be categorised into two types:

- data, which is pure information processed (added to, subtracted from, multiplied, printed, copied etc.) by a module

- control data, which causes a module to take different actions dependent upon the value of the data.

Parameters consisting of control data are always superfluous and can be removed. For this reason, control data parameters result in higher degrees of coupling. Control data parameters can be removed by the control decision being made at higher level in the module hierarchy, as shown in figure 7.12.

Binding Time. Binding refers to the time at which data references between modules become fixed and this can occur at various times during the development of software with later binding times leading to a lower degree of coupling. For example, if two modules can be independently compiled they will be easier to maintain than if they must be compiled as a single unit.

In addition to a straightforward analysis of the interface between two modules, an increase in the degree of coupling can also arise when two or more modules access a common environment. Common environments occur where some data repository exists such as a file, database or common data area.

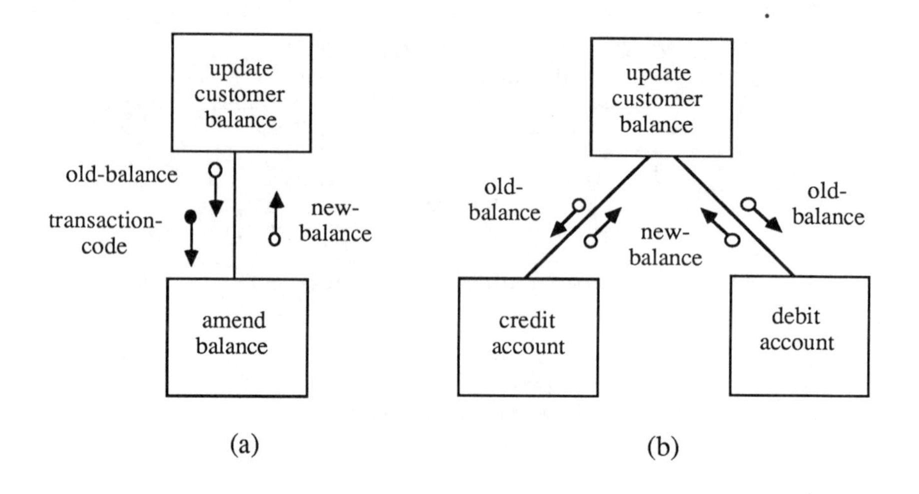

Figure 7.12: Removing Control Flows

Consider, for example, the file in figure 7.13. Module A is responsible for the placement of data in the file, whilst module B reads this data. Should at some stage the manner or type of data which is written by module A change, such as the addition of a field to each record in the file or a change in the manner in which the file is organised, this will have an immediate impact on module B which will also require amendment in order to reflect the new way in which module A writes data to the common file.

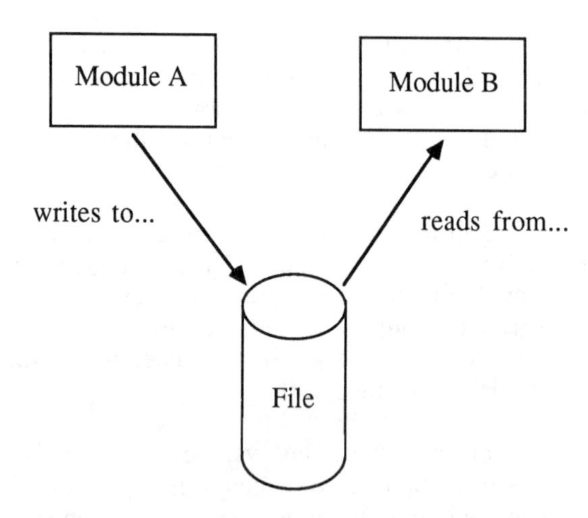

Figure 7.13: Common Environment Coupling

Some programming languages, such as COBOL, attempt to reduce the problems associated with this type of coupling by providing compile-time facilities for copying-in separate file definitions contained in a special definitions library. However, such facilities still require the recompilation of programs which access the changed files.

Because of its subtle nature, common environment coupling is regarded as the highest degree of coupling, whilst data coupling is regarded as the lowest degree. Figure 7.14 shows how the respective types of coupling relate to each other.

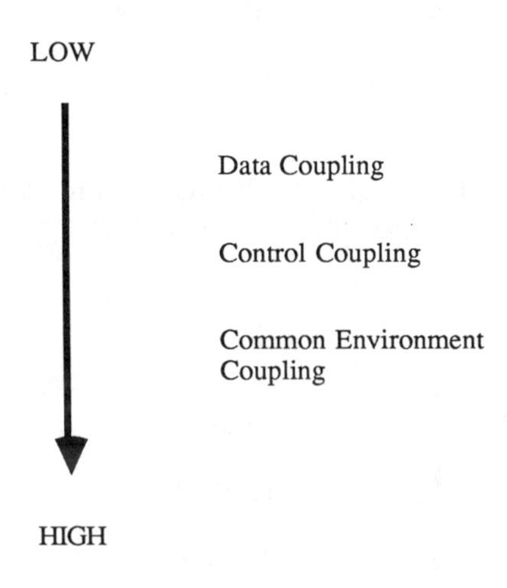

Figure 7.14: Degrees of Coupling

Decoupling is achieved by the process of making modules less dependent on each other through the reduction of common environments and the simplification of complex interfaces. In particular, the following practices are recommended:

- removal of common data areas
- documentation of remaining common data areas, such as files, clearly identifying modules which write to the area and modules which read the data
- change parameters for the latest possible binding
- removal of control data parameters.

7.2.2 Cohesion

The concept of cohesion relates to the degree to which a module performs one and only one task and like coupling, varying degrees of cohesion are recognised. High cohesion is represented by modules which have precisely one task to perform, such as to update a customer's credit balance or to produce a despatch docket. Highly cohesive modules are often referred to as having a single-minded purpose, i.e. one job to do. Poor cohesion, on the other hand, is represented by modules which perform either many tasks or only part of a task. A multi-task module may accept sales data, produce a despatch docket, calculate a salesman's commission, update the customer's credit balance and produce a sales receipt. The problem with poor cohesion is that relatively unrelated tasks are placed in the same module, thus causing difficulties in maintenance.

As with coupling, the degree of cohesion within a module may take varying degrees and these shown in figure 7.15. The following degrees of cohesion are recognised.

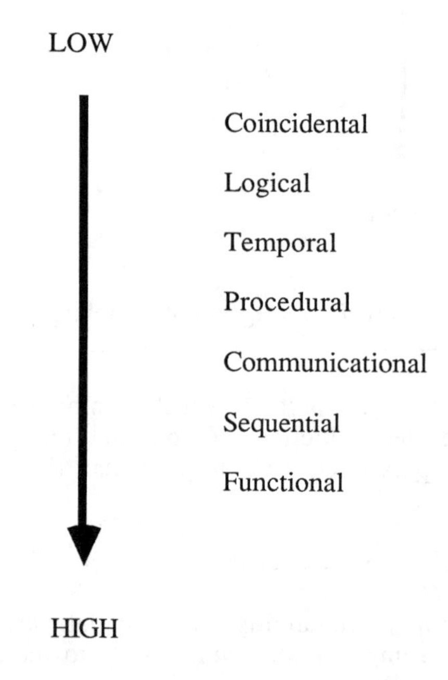

Figure 7.15: Degrees of Cohesion

Coincidental. Coincidental cohesion arises, more often than not, from code already written and which has been divided up into so-called 'manageable units'. Unfortunately, the practice of writing a monolithic coded structure and, as an afterthought, dividing it up into modules, usually only causes more problems than it solves, since each unit of code is associated purely by coincidence and has no logical function which as an independent unit it fulfils.

Logical. Logical cohesion refers to those modules which perform a set of related functions. For example, a module called DATA-INPUT, may be a general data acceptance routine which acts as a filter to all data input. The module may display standard VDU screens and perform data validation checks on check digits, number ranges and data values. This type of cohesion has some grounds for acceptability in so much as it attempts to construct modules which have some related function, even though they are often general-purpose, containing many different control paths.

An example of logical cohesion is shown in figure 7.16 in which two modules, C and D, both call a common error-handling routine called ERROR. The ERROR module is passed a control parameter indicating an error number which is used as an index to obtain an error message text in the ERROR-TEXT file. The module ERROR is thus an example of logical cohesion since it handles all references to the file ERROR-TEXT. The practical implications for system maintenance are that if the physical format or structure of the file ERROR-TEXT is changed, only the module ERROR will require amendment.

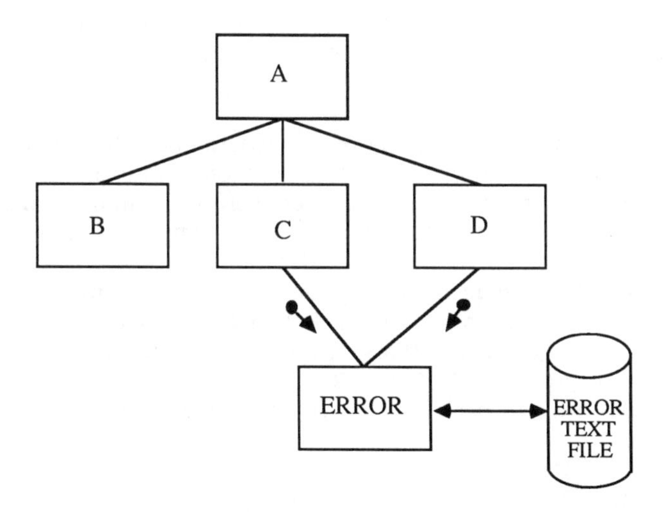

Figure 7.16: Example of Logical Cohesion

Temporal. Temporal cohesion relates to modules which contain a set of actions, each of which may be independent of each other, but all take place within a limited timespan. For example, a module may initialise all the data structures used by a suite of programs. Such cohesion is acceptable because the activities are restricted to a particular timespan but, it also has the disadvantage of often isolating important initialisation and termination activities from the main bodies of system functions.

Procedural. Procedural cohesion relates to activities which must be performed by a system, often as an iteration. Consider, for example, a system which accepts a series of sales returns before performing an analysis. If the code concerned with the acceptance of the sales returns is performed by a single module which is repeatedly called, the module is regarded as having procedural cohesion.

Communicational. Communicational cohesion applies to those modules which typically perform a range of functions on a single data structure. For example, a module may perform all screen handling or file accessing within a given system. Communicational cohesion marks the beginning of truly desirable cohesion since it attempts a sensible decomposition of a system through grouping together activities which all access a single object. The effect of this is that if the characteristics of the object are changed, such as a new screen size or file organisation, only a single module has to be amended. Communicational cohesion is therefore often found in software which is to be ported between machines, machine-dependent operations such as file handling all being placed in a single module.

Sequential. Sequential cohesion occurs where one action produces a data output which is the input to the next action. For example, a module in which statements obtain a data value, perform arithmetic manipulations on it and rewrites, would be regarded as having sequential cohesion.

An example of sequential cohesion is shown in figure 7.17 in which a structure chart has been derived from a simple data flow diagram. The data flow diagram represents a series of data transformations from a to d by the processes P, Q and R and the sequence of corresponding modules are sequentially cohesive, since one receives the output of the other.

Functional. Functional cohesion is the strongest form of cohesion and arises where a module achieves a single function, such as 'find the square root of a value', which cannot sensibly be further decomposed.

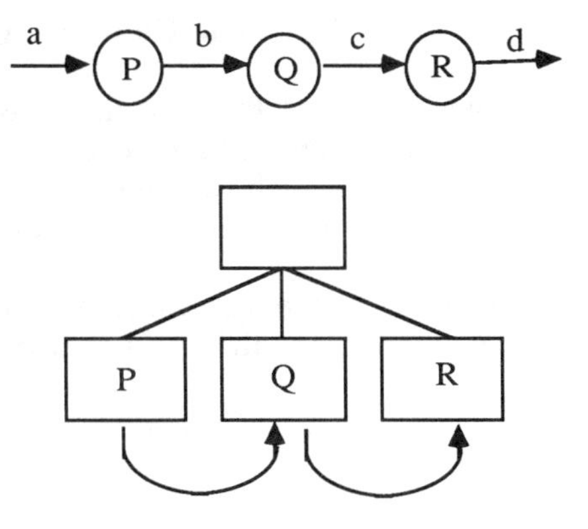

Figure 7.17: An Example of Sequential Cohesion

7.2.3 Summary

In this section, it has been shown that good system module design requires that modules have a high degree of independence. This independence is expressed in terms of *loose coupling*, in which modules have simple interfaces and a low occurrence of commonly accessed objects and *high cohesion*, in which modules have a single-minded function. In the following section, we examine the mechanism by which a good module design is derived from a system specification.

7.3 Process-Driven Design

This section discusses the steps by which a basic design structure is derived from a process specification. It is assumed that this specification has been expressed as a data flow diagram. Three steps are involved:

- identification of the characteristics of the data flow diagram
- employing the techniques of transform or transaction analysis to produce a structure chart, depending upon the characteristics identified in the data flow diagram, and
- optimisation of the structure chart using design heuristics to produce highly cohesive, loosely coupled modules.

The result of following these three basic steps is the production of a well-structured design, upon which a detailed design can be built. These three steps are now outlined in greater detail.

7.3.1 Identifying Data Flow Diagram Characteristics

Figure 7.18 shows sections of two data flow diagrams and the briefest observation reveals that each diagram has essentially a different structure. In diagram (a), the process X appears to lie at the centre of the diagram with all flows either converging on the process or flowing away from the process.

In diagram (b), on the other hand, process Y, which also plays a central role, appears to have a single input data flow, the result of which generates one or more output data flows.

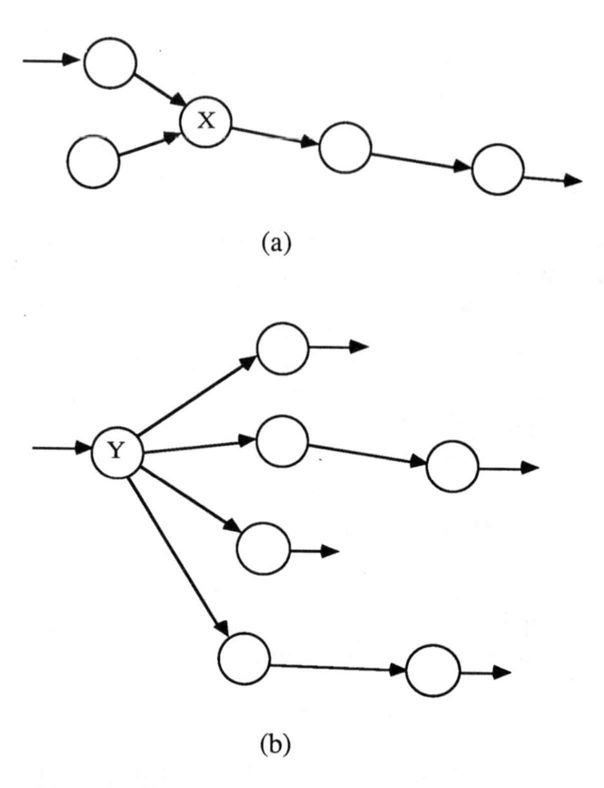

Figure 7.18: Different Data Flow Structures

These two data flow structures are commonly occurring themes in larger data flow diagrams and represent an important categorisation of systems. The structure shown in diagram (a) consists of a system which appears to be concerned, initially with the collection of data, followed by its transformation in process X and finally its distribution. Consequently, process X is known as the *central transform*. In diagram (b) however, process Y triggers one or more alternative data flow paths depending upon the input data to the process and hence process Y is known as a *transaction centre.*

Thus the first stage in deriving a design structure is the identification of whether a data flow diagram is *transform-based* or *transaction-based*, the next step being dependent upon which of these characteristics the data flow diagram exhibits. In the case of a transform-based system, *transform analysis* is employed, whilst a transaction-based system requires the use of *transaction analysis.*

7.3.2 Producing a Structure Chart

Transform Analysis

The basis of transform analysis is the identification of those processes which perform the basic functions of a system, but without reference to specific input or output procedures. These procedures are known as the central transform and as such represent the centre of a system.

In diagrams such as figure 7.18(a), the identification of the central transform is trivial, however, much larger systems require a systematic approach to its identification. This is achieved by analysing the types of data flows within a system. Two main types of data flow may be identified.

Afferent Data Flows. Afferent data flows are data flows which pipe data between processes which are concerned with the input of data to the system from external sources. The broken line in figure 7.19 shows the boundary of afferent data flows in a sample data flow diagram. All data flows above the broken line are concerned with obtaining data either directly from users, from files, or by computation and feed into the process *produce guest's bill.*

Efferent Data Flows. Efferent data flows are data flows which pipe data between processes concerned with the distribution of data from the central functions of the system to external sources. For example, all data flows below the broken line in figure 7.20 are efferent data flows since they link processes concerned with the post-processing of data.

The central transform of a system consists of those processes which lie neither within the afferent data flow boundary nor within the efferent data flow boundary. Often different analysts will choose slightly different afferent and efferent data flow boundaries since the inclusion of any given process is normally a matter of judgement.

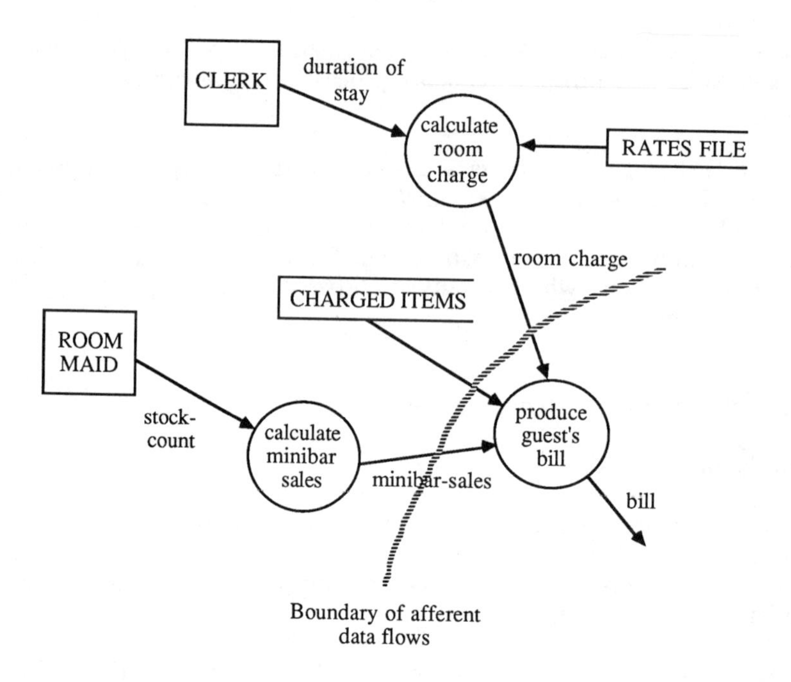

Figure 7.19: Afferent Data Flows

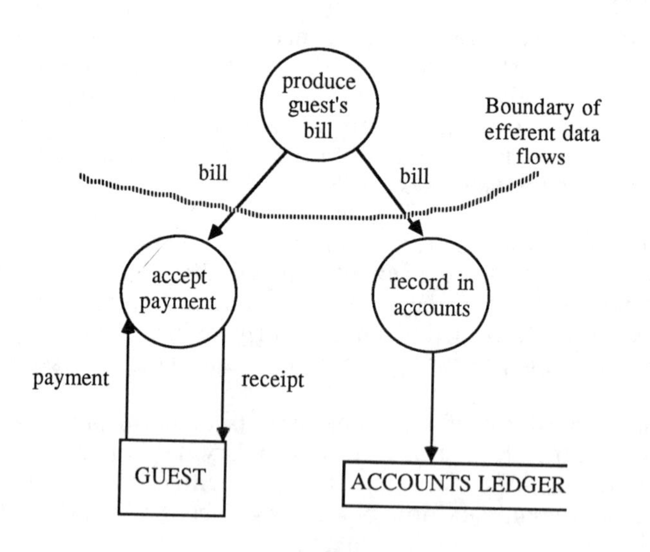

Figure 7.20: Efferent Data Flows

The important point is that it is generally better to chose any reasonable boundary and to continue with the design process, than to perform numerous iterations attempting to get a boundary exactly right.

Once the central transform has been identified, it is possible to produce a first-cut structure chart.

The upper half of figure 7.21 shows a sample data flow diagram together with its afferent and efferent data flows and central transform and the lower half of the figure shows how this basic structure maps onto a structure chart.

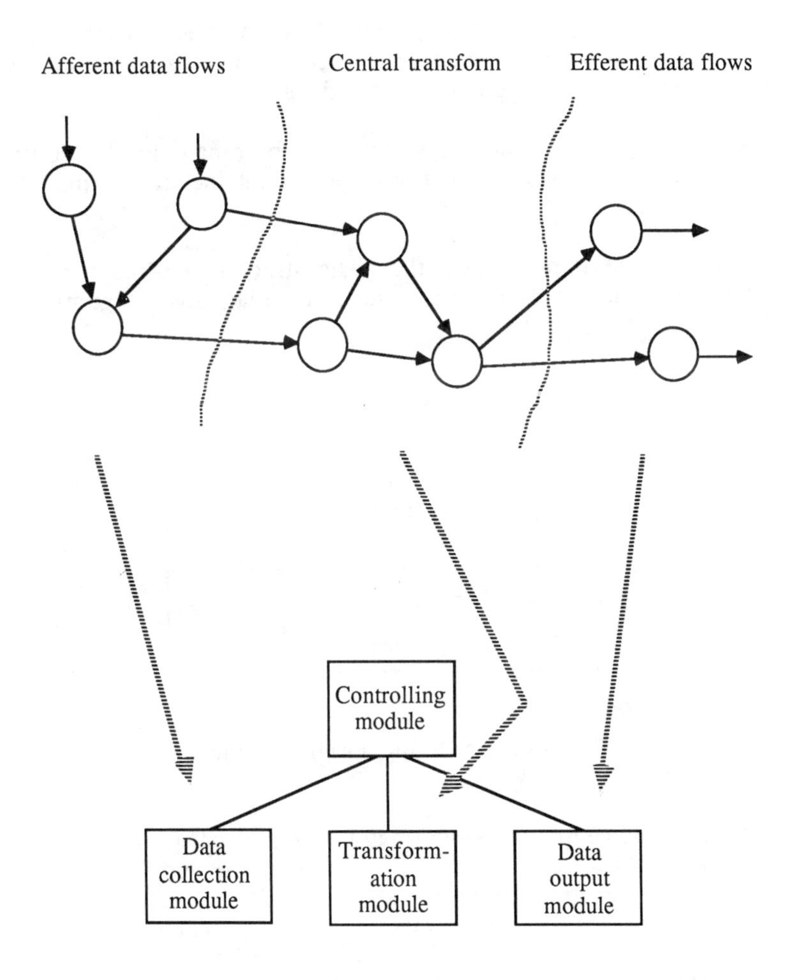

Figure 7.21: Transform Analysis

The steps by which the structure chart is produced are through the definition of the following modules:

- a controlling module, whose function is to control the basic system functions of obtaining data, the transformation of data and the output of the transformed data

- a data collection module, which is subordinate to the controlling module and whose function is to co-ordinate the receipt of all data to the basic system functions

- a transformation module, which is subordinate to the controlling module and whose function is to co-ordinate and perform the basic system functions upon the received data

- a data output module, which is subordinate to the controlling module and whose function is to output the transformed data as required.

Using these general principles, the basic structure outline for the hotel check-out subsystem, based upon the data flow diagrams in figures 7.19 and 7.20, is shown in figure 7.22.

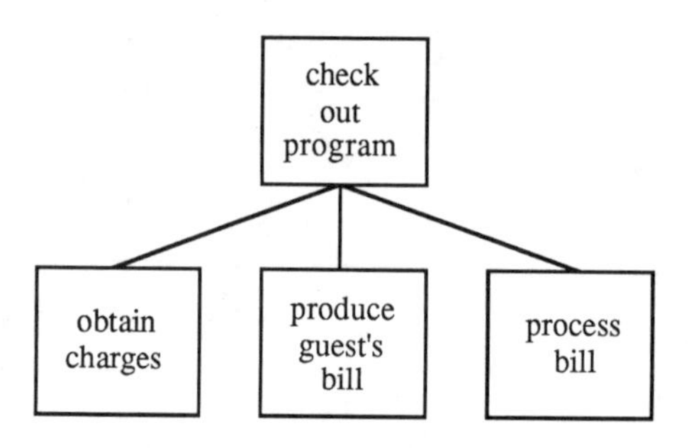

Figure 7.22: Outline Program Structure

In some cases, no central transform may be obvious. This situation normally arises where a system consists of a series of related but independent subsystems. Under these circumstances, the basic structure shown in figure 7.22 is modified, to that shown in figure 7.23, so that subordinate to the controlling module is a series of subsystem controllers, each of which have their own data collection, transformation and data output modules.

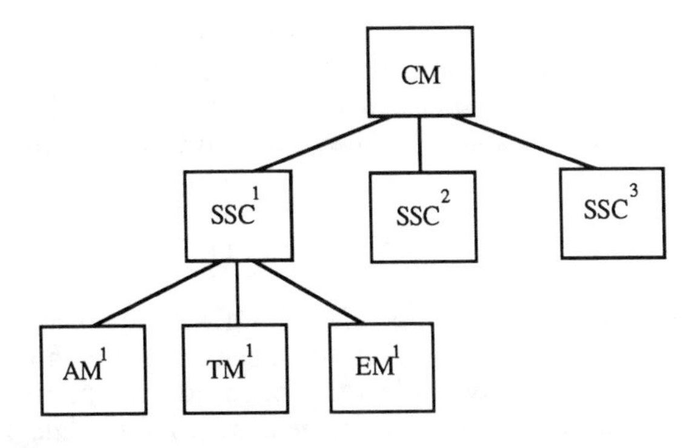

Figure 7.23: Transform Analysis with Subsystems

Having achieved a first-cut structure chart, it is possible to expand the structure chart into a second-cut through detailed specification of the afferent, efferent and transformation branches in turn.

Beginning with the afferent branch, the next level of modules is produced, which will be subordinate to the data collection module. Often, these modules will correspond to processes within the afferent portion of the data flow diagram, but it is possible for several of these processes to be combined to form a single module. Each module should be labelled with a name representative of the module's function. Figure 7.24 shows the second-cut program structure for the hotel check-out program.

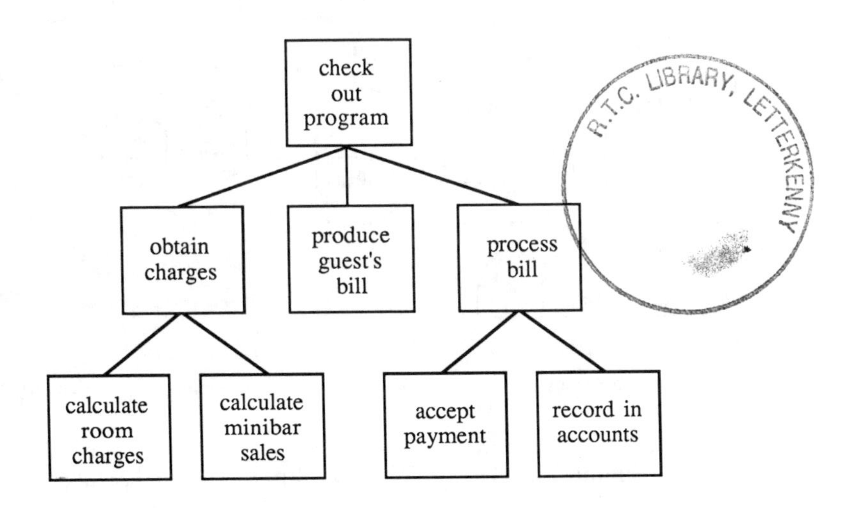

Figure 7.24: Second-Cut Program Structure

If warranted, further levels of modules should be specified until the lowest level module achieves functional cohesion, namely that it performs one and only one function. During this process, modules may also be removed, where appropriate. For example, figure 7.25, which shows the final version of the hotel check-out program, has lost the module *process bill*, as this had no real function. Instead, the controlling module (*check out program*) now calls the modules *accept payment* and *record in accounts* directly.

Finally, the structure chart should be elaborated with the parameters between modules and further annotated with file accessing and human-computer interaction. A variety of symbols may be used, but in figure 7.25, the symbols below the modules *calculate room charges, calculate minibar sales* and *accept payment* all show human-computer interaction; whilst the symbols below the modules *produce guest's bill* and *record in accounts* represent file accessing.

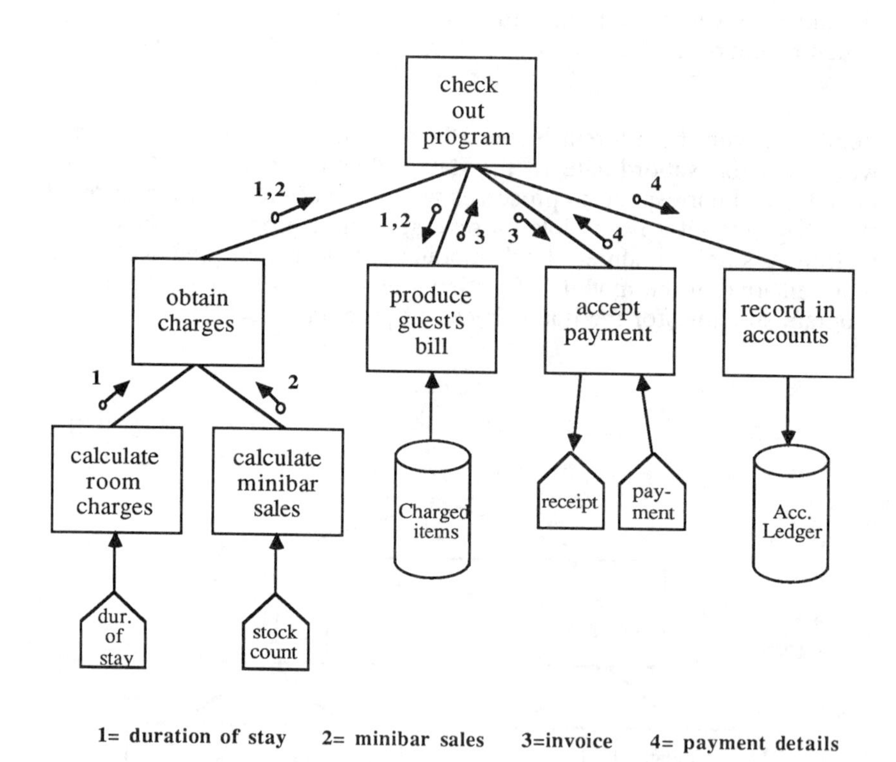

Figure 7.25: Hotel Check-Out Program- Final Structure

Transaction Analysis

The basis of transaction analysis is the identification of the process whose incoming data triggers one or more other processes. The data flow into the transaction centre is known as the reception path and the resulting data flows are known as action paths. Like transform analysis, these paths must first be identified. Fortunately, recognition of a transaction centre is usually relatively simple as a result of the radial nature of their data flows.

Once the transaction centre has been identified, a first-cut structure chart is produced, based upon the general structure shown in figure 7.26. The process *T* in figure 7.26 represents the transaction centre.

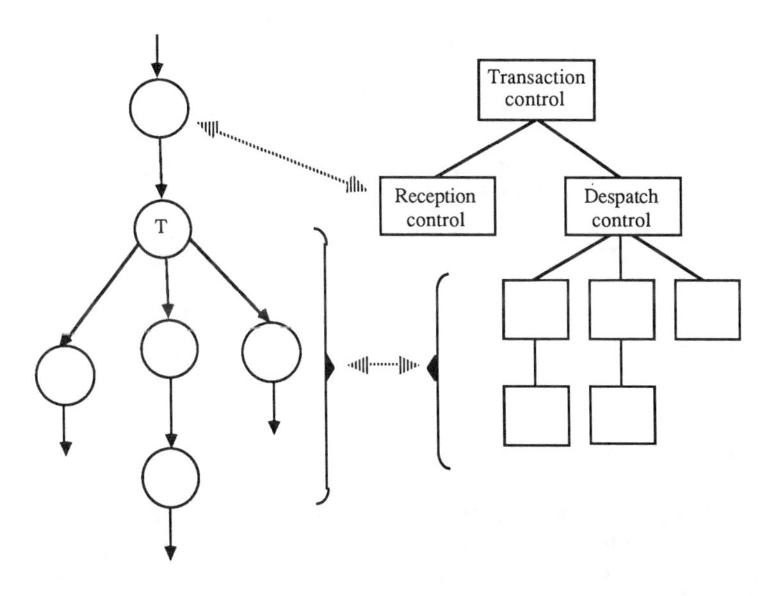

Figure 7.26: Transaction Analysis

On the corresponding structure chart, three initial modules are required:

- a transaction control module,which is responsible for the overall co-ordination of the transaction processing

- a reception control module, which is subordinate to the transaction control module and whose function is to control the receipt of transaction details

- a transaction despatch module, which is subordinate to the transaction control module and whose function it is to control the activation of transaction processing routines based upon the transaction details received from the reception module.

The lower level modules of the reception and despatch modules are subsequently detailed and refined.

7.3.3 Optimising the Structure using Design Heuristics

Once a basic structure chart has been obtained by either transform or transaction analysis, a variety of heuristic design techniques, based upon module independence, can be applied in order to improve upon the design. Typical of these techniques are the following:

- explosion or implosion of modules to achieve sensible factoring

- ensuring minimum coupling between modules and maximum cohesion within modules

- removal of data collection modules where unnecessary.

In addition, structure charts should be labelled with the primary data flows and control parameters that are passed between calling and called modules.

7.4 Data-Driven Design

An alternative approach to module design is through the derivation of a module structure based upon the data structures to be processed. This approach has received wide acceptance and is largely based upon the work of Michael Jackson and his programming method known as Jackson Structured Programming or JSP (*Jackson, 1975*). We shall use the term JSP forthwith to refer to the technique of data-driven module design.

7.4.1 Overview of the Technique

The underlying philosophy of the JSP technique is based upon the recognition that the structure of a module should have some rational foundation and because programs manipulate data, that foundation should be based upon the data to be processed. The philosophy of the technique therefore, is to generate program structures based upon the merging of input and output data structures. The technique consists of four basic steps.

Record Data Structures. The first step consists of recording the input and output data structures to be processed. A variety of notations may be used to record these structures, such as the data specification language introduced in section 4.4.1. However, a more popular graphical notation is the use of data-driven structure chart notation in which data items are decomposed into elementary components. Standard sequencing, repetition and selection notation can also be applied. Figure 7.27 shows examples of data structure diagrams.

Compose a Program Structure From Data Structures. Once the input and output data structures have been specified, a technique of identifying correspondences and merging of data structures can be applied in order to derive a program structure. A correspondence is is a direct mapping between an input and output data component.

List and Assign Elementary Operations. Having obtained an outline program structure, the next step is to list the elementary operations which will perform the required transformations between input and output data structures and assign these to modules. The result will be an elaborated program structure.

Translate into a Programming Language. The final step is to convert the elaborated program structure into a programming language representation.

7.4.2 Example

An example of the way in which the JSP method may be applied is to consider the hotel case study. Suppose the system maintains a file of transactions on guests' accounts, sorted by room. It records charges made to the account, such as bar drinks, restaurant meals etc. and payments made by the guest. Assume also that the hotel requires a formatted output report for the contents of this file. Figure 7.27 shows the basic data structures of the transaction file, *trans-file*, and the output report, *trans-report* and a sample of the required output is shown in figure 7.28.

The *trans-file* structure shows that the file consists of an iteration of *guest-groups*, each of which consist of an initial balance (the amount carried forward) and a group of transactions. Each *transaction-group* consists of transaction items (*trans-item*), which may be either a *receipt* or a *charge*. These in turn could be further elaborated to show room number and amount. Similarly, the *trans-report* structure shows that the report consists of *headings* and a *report body*. The body in turn consists of a breakdown of each guest's details, including the initial balance, transaction details and revised current balance.

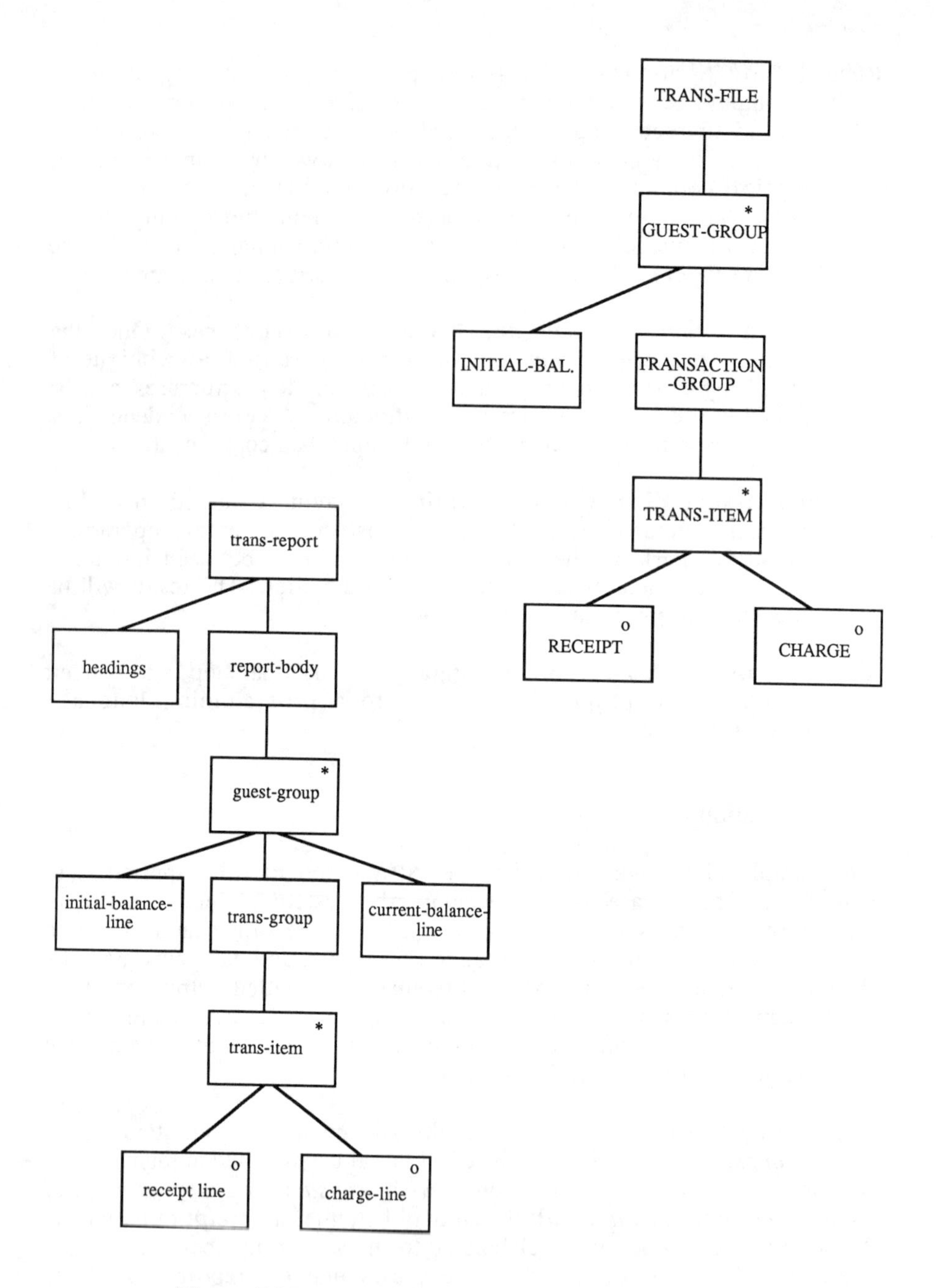

Figure 7.27: Data Structure for the Hotel Case Study

```
HOTEL ADMINISTRATION SYSTEM                 9 JUL 88   10:13    Page  1

                       Guest Account Transactions

Room  Guest's name   Arrival  Credit        Transaction Details
                              Limit       Date    Type         Amount

101   Jones          7.7.88   2000.00   7.7.88    B/f            0.00
                                        7.7.88    Room          52.00
                                        8.7.88    Breakfast      6.50
                                        8.7.88    Room          52.00
                                        8.7.88    Paid         162.50 -

                                                  Total          0.00
103    Wilson        8.7.88   2000.00   8.7.88    B/f          133.23
                                        8.7.88    Restaurant    47.38
                                        8.7.88    Bar            7.52

                                        continued
```

```
HOTEL ADMINISTRATION SYSTEM                 9 JUL 88   10:13    Page  2

                       Guest Account Transactions

Room  Guest's name   Arrival  Credit        Transaction Details
                              Limit       Date    Type         Amount

103    Wilson        8.7.88   2000.00   9.7.88    Room          52.00
                                        9.7.88    Restaurant    13.92

                                                  Total        244.05
```

Figure 7.28: Sample of Required Output Report

Having established the input and output data structures, the next step is to identify the correspondences between the components in each data structure. In this example, the correspondences are between:

TRANS-FILE - trans-report
INITIAL-BAL. - initial-balance-line
TRANS-ITEM - trans-item
CHARGE - charge-line
GUEST-GROUP - guest-group
TRANSACTION-GROUP - trans-group
RECEIPT - receipt-line

Using these correspondences, the input and output data structures can be merged, as shown in figure 7.29, to derive an outline program structure.

The final step of the design process is to assign elementary operations to the program structure and in this example they would consist of:

1. open files
2. write heading print lines
3. read first initial balance record
4. write initial balance print line
5. assign initial balance to current balance
6. read first transaction item
7. write transaction print line
8. subtract payment from current balance
9. add charge to current balance
10. read next transaction
11. write current balance print line
12. close files

Figure 7.30 shows the completed program structure for the sample application. Notice that each module has been assigned an appropriate process name and the elementary operations shown above have been attached to the processes.

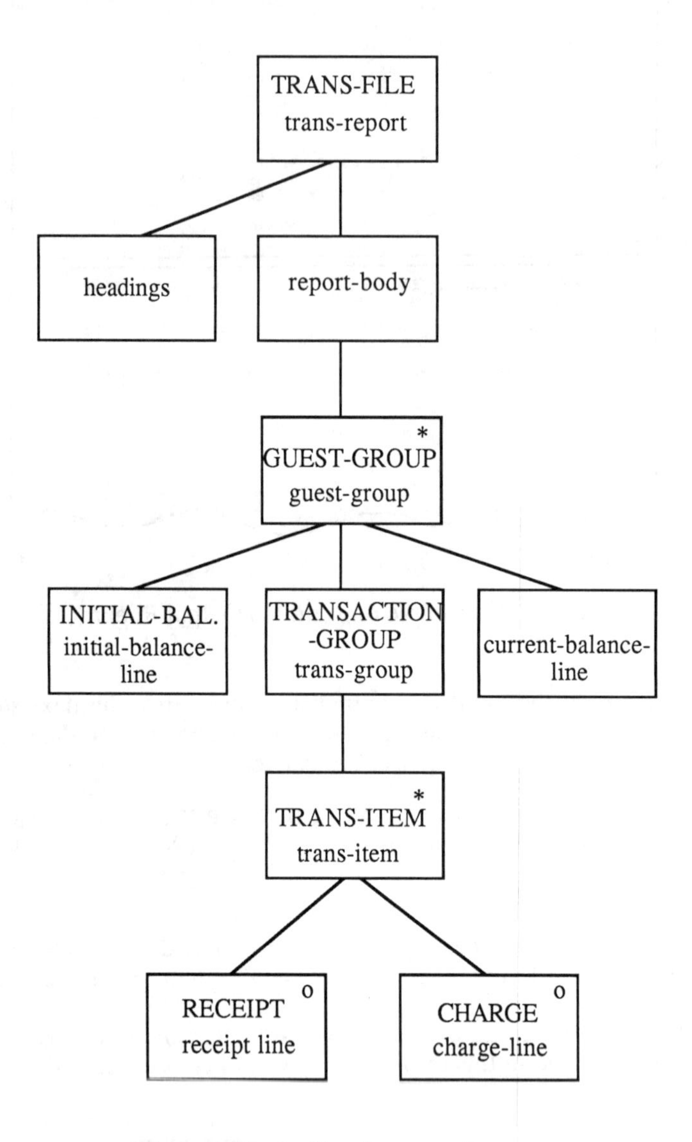

Figure 7.29: Outline Program Structure

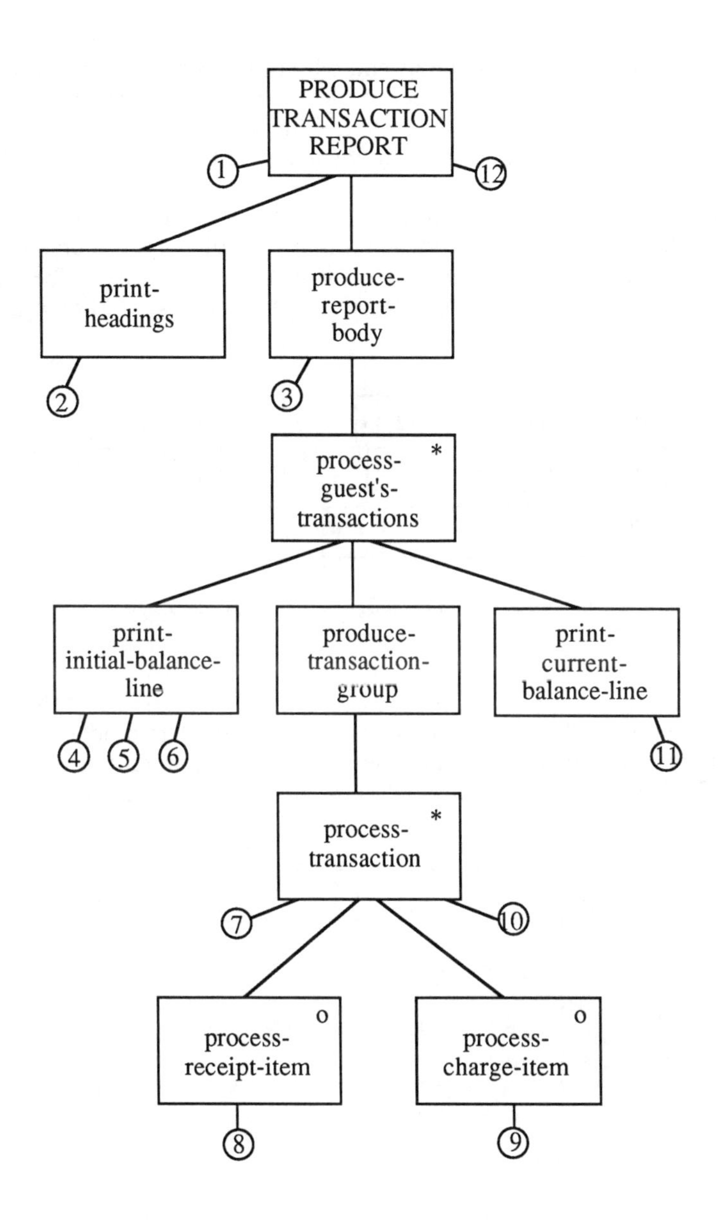

Figure 7.30: The Final Program Structure for Transaction Reporting

7.4.3 Boundary Clash

In the previous example, the input and output data structures were such that there was a neat fit between them and it is a relatively simple operation to derive a program structure. However, data structures often do not fit so well and under these circumstances a *boundary clash* is said to occur.

Consider for example, a system in which data is to be read into a program from an electronic sales till. Assume that the till outputs details of each transaction, sorted into a product code order. For reasons beyond the system developer's control, the data is output in blocks, as shown in figure 7.31.

001	12.34	45.33	22.33	1.57	***	002	1.17
23.47	19.23	***	003	9.18	17.46	53.57	***
004	19.55	2.33	***	***			

Figure 7.31: A Sample Till Output

The shaded areas represent product codes and the asterisks (***) act as a marker to indicate the end of the transactions for the last product code. The two sets of asterisks at the end of the data represent an end of data marker. Figure 7.32 shows the summary output required from the proposed system.

```
RESTAURANT SALES TILL RECORD

Product code        Sales value

   001                 81.57
   002                 43.87
   003                 80.21
   004                 21.88
```

Figure 7.32: A Sales Summary Report

At first a structure clash is not apparent, however, as soon as the respective data structures are drawn, as in figure 7.33, it can be seen that apart from the fact there is one set of till details and one sales summary, there are no other correspondences in the data structure.

Furthermore, careful consideration will reveal that the two structures cannot be merged to derive a program structure. This is because the till sees the data it outputs as a series of blocks, whilst the proposed summary program requires to see the data as a set of group entries.

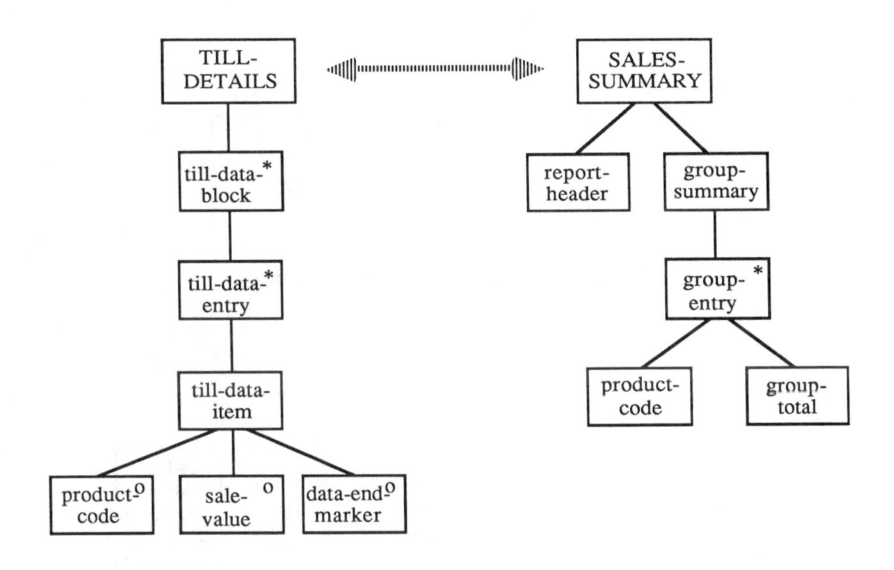

Figure 7.33: Till Output and Sales Summary Report Data Structures

At this point, the temptation is to *fiddle* the data structures to force correspondences. However, this should be avoided as it will lead to artificial program structures.

The solution is to introduce a process whereby an intermediate data structure is created which will allow correspondences to be achieved. This strategy, in which an intermediate data file must be created, is outlined in figure 7.34. With this approach, two programs are required. The first program, shown as *unpack-program* in figure 7.34, reads the blocks of data generated by the till and separates each *product-code* and *sales-value* into separate entries. The second program, shown as *summary-program* can then read the intermediate file and pick up the required data in the correct form. Figure 7.35 shows the correspondences between the three data structures.

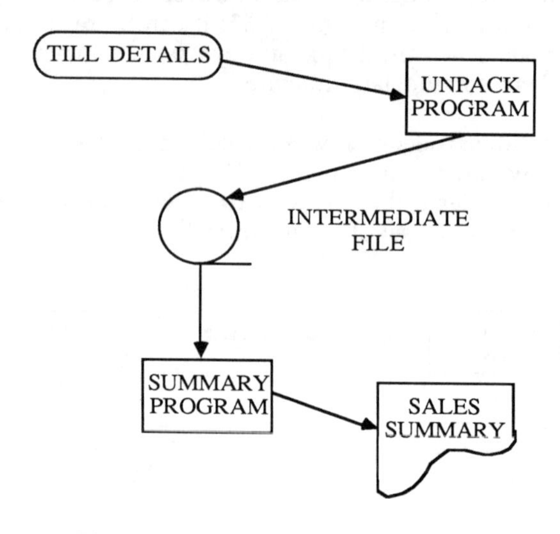

Figure 7.34: Implementation Strategy

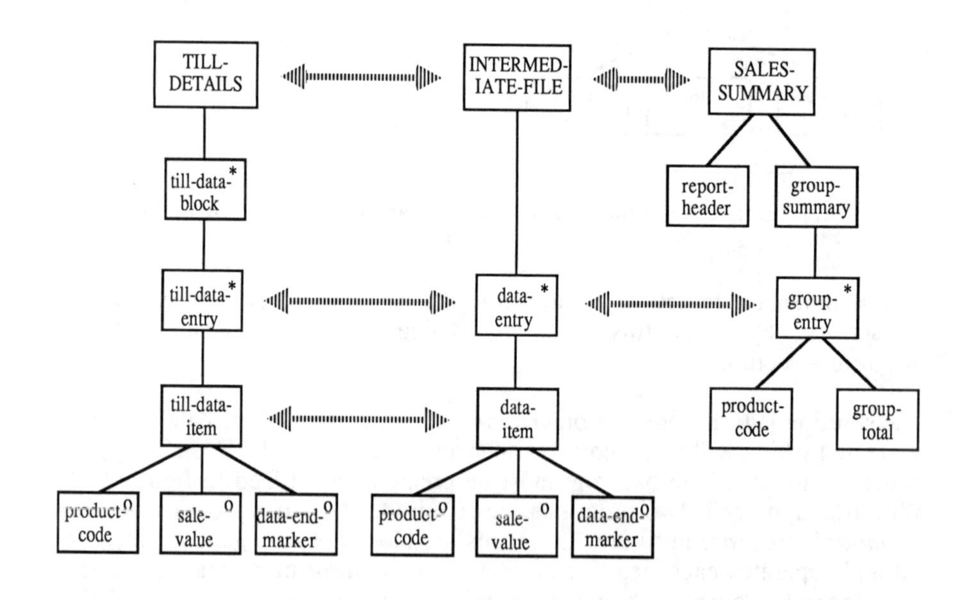

Figure 7.35: Intermediate File Data Structure

This process of creating an intermediate file is known as *inversion* and is a general technique which can be used for other types of data structure clash.

When implementing such inverted systems, it is possible to eliminate the intermediate file by making one of the two programs call the other (i.e. *invert* one program with respect to the other).

7.5 Summary

The initial task of a system designer is to define the basic structure of a system, based upon the specification for that system. In attempting this task, the overriding factor is to ensure that each module within the system has a high degree of independence, in order to ease the problems of future maintenance of the system.

Module independence is embodied in the concepts of cohesion and coupling and is achieved either through the process-driven techniques of transform and transaction analysis or through the data-driven techniques of data structure correspondences.

Further reading on process-driven design can be found in *Yourdon and Constantine* (*1975*) and *Myers* (*1975*) and for data-driven design in *Jackson* (*1975*).

Chapter 8

Data Design

Organisations collect and process large volumes of data in order to satisfy their operational needs. This data is stored on backing storage devices, such as magnetic tapes and magnetic discs. In order that data may be stored for effective and efficient processing, a developer must consider how data is to be organised in backing storage devices.

In chapter 5, the concepts which represent the logical view of data during the analysis phase of system development were discussed in the context of the conceptual schema. In contrast, concepts such as records and data items represent the physical representation of data and it is the purpose of this chapter to examine how particular file, record and data item structures are derived for any given system.

This chapter begins with a discussion of the issues associated with the design of files and databases. It is assumed that the reader has a knowledge of file structures, database models and database management systems. The chapter examines two techniques for designing record structures: *data normalisation* and *direct mapping from a conceptual schema.* Finally, there is a discussion about the influences on the design of data from the implementation perspective for both files and databases.

Design of files is considered in terms of accessing and processing requirements and the file organisation strategies available to a developer for satisfying these requirements. Database design requirements are introduced within a framework of the ANSI/SPARC architecture. The difference between the various database models is discussed in terms of differences in expressing relationships and design aspects are considered in terms of *schema design.*

8.1 Overview

Stored data lies at the heart of any data processing system. *Records* are one of the fundamental concepts of any information system and are found in a variety of forms: computer storage, card indices, filing cabinets etc. A record is used to hold data about a particular object within an organisation. For example, a record about customers might contain information on customer number, their name and address and the value of sales to the customer. These items of information are known as data items and represent the constituent parts of a record.

8.1.1 Data Design

Data design refers to the process of constructing record structures and organising these structures in such a way that their accessing is carried out in an efficient way; whilst efficiency refers both to the time taken to access a record and to storage considerations of all the records.

Data design requires input from two major development activities, as shown in figure 8.1, process modelling and conceptual data modelling.

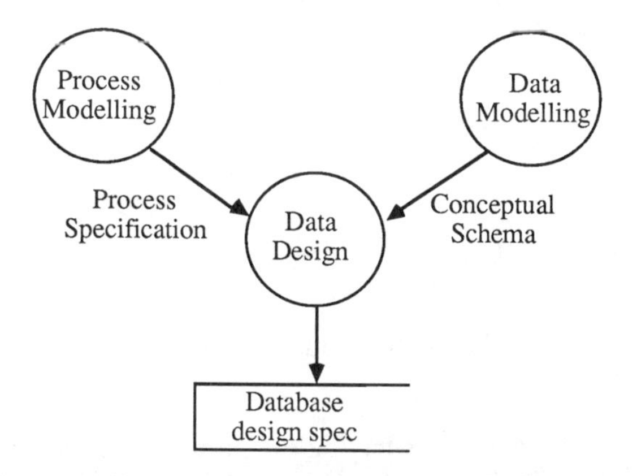

Figure 8.1: A View of Data Design

Data analysis, as discussed in chapter 5, provides the necessary knowledge about the objects which need to be stored, including knowledge about relationships between these objects and static constraints applying on them. Process analysis and process design, as discussed in chapters 4 and 7 respectively, provide essential information about the *transactions* which will manipulate the data.

8.1.2 Transactions and Data

The behaviour of an automated information system is determined by a collection of *transactions* which are performed by the system (often concurrently). The relationship between data and transactions is shown in figure 8.2.

Manipulation of the data is achieved through a number of transactions which correspond to the processing requirements of the system. Following the generally accepted definition of transactions c.f. *(Gray, 1981; Sakai, 1981; Borgida, Mylopoulos & Wong, 1982)* a transaction is defined as the set of atomic operations which when they have been executed leave a database in a consistent state, as defined by a set of constraints. A set of atomic operations (get, modify, insert, delete etc.) constitutes a logical unit of work.

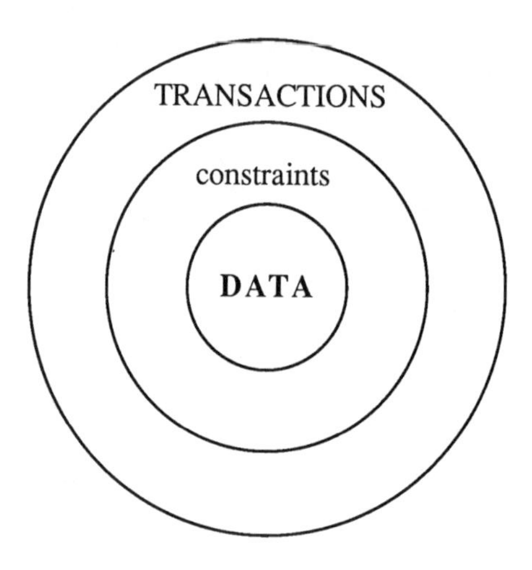

Figure 8.2: Transactions and Data

Transactions are regarded as having the properties of consistency, atomicity and durability. The term *consistency* is used to refer to the concept that a transaction must obey a set of predefined rules. Whilst *atomicity* and *durability* are used for the concepts that either all actions are done and the transaction is said to *commit*, with its effect persisting from then on, or none of the effects of the transaction survive and the transaction is said to *abort*. This is a useful definition particularly with regard to recovery procedures (see chapter 6).

8.2 Record Design

The starting point for designing the structure of a record should be a conceptual schema. In practice two approaches are favoured: using the data normalisation technique and mapping a conceptual schema directly onto record structures. The first technique requires an analysis of *functional dependencies* on attribute types and is normally carried out in conjunction with entity-relationship modelling. The second case can be carried out in a reasonably straight forward way assumming that the fact-based model has been used for the derivation of the conceptual schema. Both techniques are described in this section.

8.2.1 Data Normalisation

Data normalisation is a step-by-step process for analysing an entity type into its constituent entity types. At each step the derived entity types exhibit certain properties which are considered to be desirable in order to avoid side effects during the operation of a database involving insertion, deletion and updating of records. In this section, four steps are considered although further refinements have been proposed (*Kent, 1983*). A key concept of the data normalisation process is that of functional dependency which is discussed next before attention is turned to the process itself.

Functional Dependency Analysis

Functional dependency analysis is concerned with the identification and modelling of relationships between attribute types. A diagrammatic notation may be used to enhance visibility of results (this is often referred to as *bubblecharting*).

Each attribute type is represented as an ellipse and each functional dependency (association) as a line between the two ellipses. The direction of the association is indicated by the direction of an arrow.

Consider for example two attribute types A and B. B is functionally dependent on A, if a value of A uniquely determines the value of B.

Examples of functional dependencies are shown in figure 8.3. In 8.3(a) B is functionally dependent on A. In 8.3(b) PART-DESCRIPTION is functionally dependent on PART-NO. This means that a value of PART-NO will uniquely identify a value of PART-DESCRIPTION. In 8.3(c) B and C are both functionally dependent on A.

In figures 8.3(a), 8.3(b) and 8.3 (c), it is assumed that the attribute type on the left of the arrow represent identifying attributes. An identifying attribute may consist of more than one element. In 8.3(d) the two attribute types X and Y are assumed to be a composite identifier. In this case, Z is functionally dependent on both X and Y. In figure 8.3(e), K is functionally dependent on part of the identifier, only on Y.

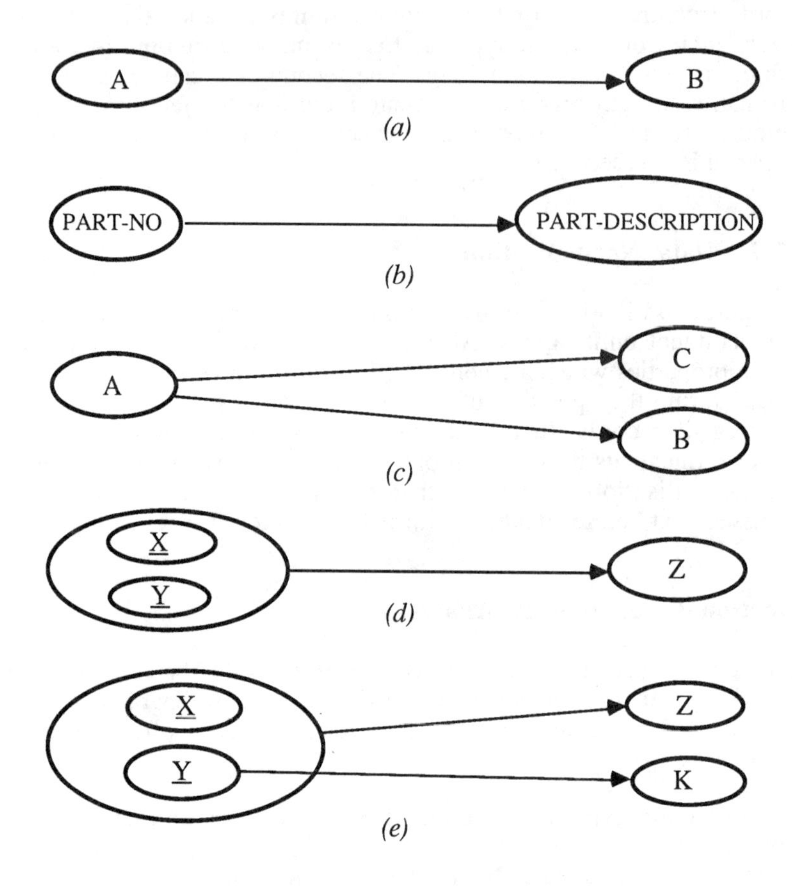

Figure 8.3: Conventions in Functional Dependency Analysis

Deriving Entity Types in Fourth Normal Form

The following steps are normally carried out during data normalisation:

- An entity type is in first normal form (1NF) if every attribute is based on a simple atomic domain i.e. there are no repeating groups of attribute types.

- An entity type is in second normal form (2NF) if it is already in 1NF and each non-identifying attribute depends fully upon the key.

- An entity type is in third normal form (3NF) if it is already in 2NF and there is no dependency between the non-identifying attribute types.

- An entity type is in fourth normal form (4NF) if it is already in 3NF and there no multivalued dependencies between its attribute types

To demonstrate these steps, consider the example of the invoice form shown in figure 8.4 and assumed that after some investigation the following entity type has been derived:

INVOICE (<u>*inv-no*</u>*, cust-no, cust-address, inv-date,*
(item-code, item-desc, item-price, quantity, price), total)

The underlined attribute type indicates an identifier.

INVOICE NUMBER:101287-300

Customer Number A12345

Customer Address 5, Highgate Hill
London, U.K.

Invoice Date : 10 / 12 / 86

Item code	Item Description	Item Price	Quantity	Price
C99	Coffee	1.00	2	2.00
B123	Breakfast	5.00	3	15.00
			TOTAL	17.00

Figure 8.4: A Sample Invoice

Deriving First Normal Forms

Following the rules for data normalisation, the first task is to remove the repeating groups. Repeating groups are rewritten as new entity types. The identifier of the original entity type must be made an attribute of each new entity (it may or may not be part of the identifier of the new entity type). In the invoice example, the repeating group is shown as the set of attribute types in the nested brackets. The result of carrying out first normal form analysis is the two entity types INVOICE and ITEM-DETAILS which are encoded as follows:

INVOICE (inv-no, cust-no, cust-address, inv-date, total)

ITEM-DETAILS (inv-no, item-code, item-desc, item-price, quantity, price)

For ITEM-DETAILS, the attribute type *item-code* is not enough to uniquely determine an occurrence of this entity. Therefore, INV-NO (the key in the original entity type) is needed for ITEM-DETAILS. The two new entity types do not have any repeating groups and therefore are in first normal form.

Deriving Second Normal Forms

The next step in the normalisation process is to analyse these two entity types to determine whether they are in 2NF. For an entity type to be in 2NF, every non-identifying attribute must be functionally dependent on the entire entity identifier.

Since there is only one attribute as part of the identifier for the INVOICE entity type there is no need to check its functional dependencies. According to the definition, this entity type is already in second normal form. The same is not true for ITEM-DETAILS and its functional dependencies are shown in figure 8.5.

It is obvious from the diagram of figure 8.5 that *quantity* and *price* are functionally dependent on the entire identifier but, *item-price* and *item-desc* are functionally dependent only on *item-code*. This means that entity type is not in second normal form. Therefore, this entity type needs to be decomposed so that all non-identifying attributes are functionally dependent on all identifying attributes. The result of this decomposition together with the unchanged entity type INVOICE is the following:

INVOICE (inv-no, cust-no, cust-address, inv-date, total)

ITEM-DETAILS (inv-no, item-code, quantity, price)

ITEM (item-code, item-desc, item-price)

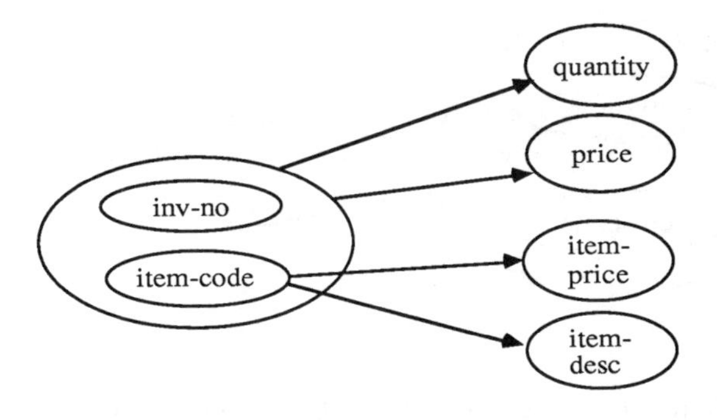

Figure 8.5: Functional Dependencies for ITEM-DETAILS

Deriving Third Normal Forms

The next step is to investigate the three entity types, to determine whether they are in 3NF. For an entity to be in 3NF, every non-identifying attribute must be independent from every other non-identifying attribute in that entity type. Figures 8.6, 8.7 and 8.8 show the results of the functional dependency analysis for the three entity types.

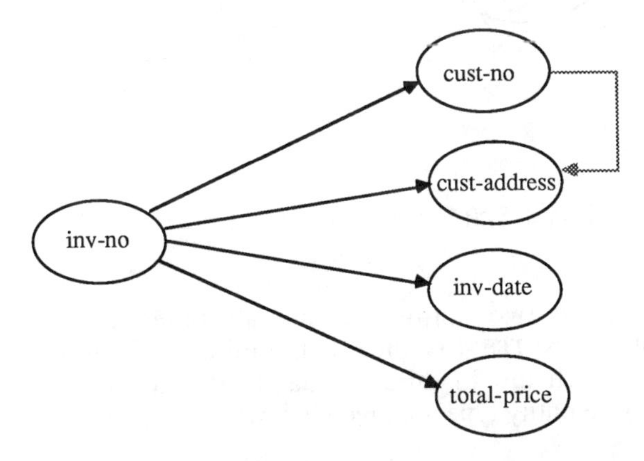

Figure 8.6: Functional Dependencies for INVOICE

For the INVOICE entity type. all non-identifying attributes are functionally dependent on the entity's identifier. But, as shown in the diagram ,the non-identifying attribute *cust-address* is functionally dependent on another non-identifying attribute, *cust-no*. Therefore, this entity type in not in third normal form and needs to be decomposed into two equivalent entity types which will be in third normal form.

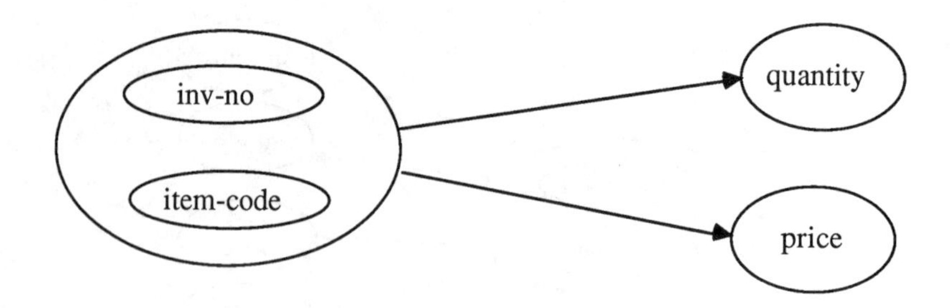

Figure 8.7: Functional Dependencies for ITEM-DETAILS

As shown in figure 8.7, ITEM-DETAILS has two non-identifying attributes and there is no dependency between them. Therefore, this entity type is already in third normal form.

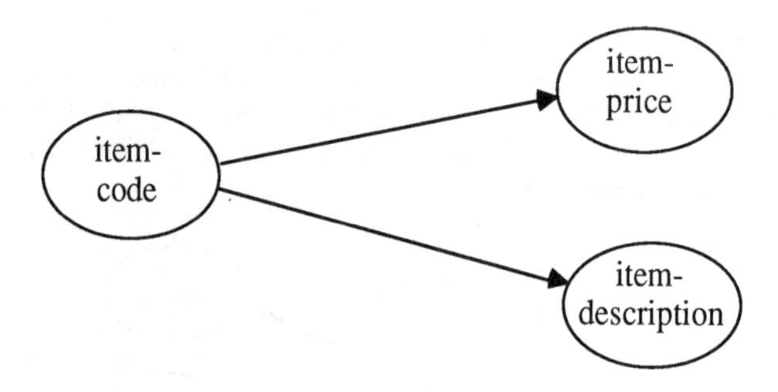

Figure 8.8: Functional Dependencies for ITEM

In figure 8.8, the two non-identifying attributes are independent and therefore entity type ITEM is already in third normal form. The full set of entity types which are in third normal form and are equivalent to the original INVOICE entity type are given below:

INVOICE (<u>inv-no</u>, cust-no, inv-date, total)

CUSTOMER (<u>cust-no</u>, cust-address)

ITEM-DETAILS (<u>inv-no, item-code</u>, quantity, price)

ITEM (<u>item-code</u>, item-desc, item-price)

Deriving Fourth Normal Forms

For many applications record design up to 3NF will suffice. However, there are some cases where, although a set of entity types may be in 3NF, multivalued dependencies may occur. Under these circumstances, a developer needs to transform the entity types to fourth normal form. Consider the case shown in figure 8.9.

AUTHOR-NO	BOOK-NO	SUBJECT	BOOK-TITLE	AUTHOR-NAME
A12	B1	Comp. Sc.	Methods	Jones
A12	B1	Maths	Method	Jones
A24	B1	Comp. Sc.	Methods	Smith
A24	B1	Maths	Methods	Smith
A12	B2	Maths	Calculus	Brown

Figure 8.9: Information about Authors and Books

By carrying out data normalisation on the information presented in figure 8.9, the following entity types in 3NF are derived:

AUTHOR (author-no, author-name)

BOOK (book-no, book-title)

BOOK-SUBJECT (author-no, book-no, subject)

This example models the fact that 'each author is associated with all the subjects under which the author's book is classified'. Consider the instances of entity type BOOK-SUBJECT shown in figure 8.10.

AUTHOR-NO	BOOK-NO	SUBJECT
A12	B1	Comp. Sc.
A12	B1	Maths
A24	B1	Comp. Sc.
A24	B1	Maths
A12	B2	Maths

Figure 8.10: Instances of BOOK-SUBJECT entity type

It can be seen that, the attribute type *subject* contains redundant values which can be deduced from other the values. The problem in this case is that the same set of subjects is associated with each author of the same book- a book has a number of authors who are all classified under the same subject matter for which the book has been classified. In this case, *Book_no* multidetermines *author-no* and *subject*. Therefore, by decomposing the BOOK-SUBJECT entity type the following set of entity types are derived which are all in fourth normal form.

AUTHOR (author-no, author-name)

BOOK (book-no, book-title)

BOOK-AUTHOR (book-no, author-no)

SUBJECT-BOOK (book-no, subject)

8.2.2 Record Design from the Conceptual Schema

The mapping from a conceptual schema to a set of entity types in third normal form is relatively straightforward, assuming that a model such a the fact-based model is used so that all label types and all relationship cardinalities are explicitly defined. This mapping has been formalised in (*De Troyer & Meersman, 1986*).

The mapping process is carried out in the following three steps.

Group Functionally Dependent Roles

To carry out this step, consider each entity type and group all functionally dependent roles. In the conceptual schema functionally dependent roles are specified in terms of the relationship cardinality of 1-1.

To demonstrate this step, consider the conceptual schema of the hotel case study in figure 5.24 of chapter 5. By following the rule defined above, the following groupings are derived (entity types are shown in upper case letter whereas label types are shown in lower case letters).

1. From RESORT
   resort_type, resort_name

2. From REQUEST
   request_no, PERIOD, RESERVATION, ROOM, CUSTOMER

3. From CUSTOMER
   cust_no, cust_name, cust_address

4. From PERIOD
   start_date, end_date

5. From RESERVATION
   res_no

6. From CHECKOUT
   checkout_no

7. From CHECKIN
   CHECKOUT, ROOM, checkin-ref-no

8. From ROOM
   room_no, HOTEL

9. From HOTEL
   hotel_no, no_of_rooms

10. From CHARGE-ITEM
    ref_no, item_desc, ROOM

11. From BILL
    bill_no, CHECKOUT

12. From BILL-LINE
    CHARGE-ITEM, BILL

13. From SUMMER-RESORT
    no_of_beaches

14. From WINTER-RESORT
    no_of_slopes

Group Related Entity Types With No Functional Dependency

This step ensures that any fact types which have not been considered in the first step are included in the set of record types.

In the conceptual schema of the hotel case study there is one fact type with no functional dependency. This is the relationship between RESORT and REQUEST. By following the rule of the second step another grouping emerges.

15. From < RESORT-REQUEST> relationship type
    RESORT, REQUEST

Substitute Entity Types By Corresponding Label Types

This step produces the final list of record types, with their corresponding data items and identifiers. During this step the following points need to be considered:

- If there is more than one label type which uniquely determines the occurrence of an entity type, then there is the case of candidate identifiers and one is chosen.

- In certain cases a combination of label types will be required to be taken as the identifier since only the value of the combined labels can uniquely determine an occurrence of the entity type (this is known as the uniqueness constraint). An example of this is the label types *start_dat*e and *end_date* which need to be taken together for identifying a particular occurrence of PERIOD.

- If there are any subtype entity types then in the grouping of each entity type (derived in the first and second steps) a reference must be added to its supertype. An example of a subtype entity type is that of SUMMER-RESORT.

Following the above rules, the resultant record types, with their identifier underlined are as follows:

1. From RESORT
   (<u>resort-name</u>, resort-type)

2. From REQUEST
   (<u>request-no</u>, start-date, end-date, res-no, room-no, cust-no)

3. From CUSTOMER
   (<u>cust-no</u>, cust-name, cust-address)

4. From PERIOD
   (<u>start-date</u>, <u>end-date</u>)

5. From RESERVATION
   (<u>res-no</u>)

6. From CHECKOUT
   (<u>checkout-no</u>)

7. From CHECKIN
   (<u>checkin-ref-no</u>, checkout-no, room-no)

8. From ROOM
   (room-no, hotel-no)

9. From HOTEL
   (hotel-no, no-of-rooms)

10. From CHARGE-ITEM
    (ref-no, item-desc, room-no)

11. From BILL
    (bill-no, checkout-no)

12. From BILL-LINE
    (bill-no, ref-no)

13. From SUMMER-RESORT
    (no-of-beaches, resort-name)

14. From WINTER-RESORT
    (no-of-slopes, resort-name)

15. From < RESORT-REQUEST> relationship type
    (resort-name, request-no)

8.3 Characteristics of File Design

8.3.1 Types of File

In order to access a record within a file, there is a need to distinguish the desired record from all other stored records. This can be achieved by selecting one or more data items as the record key, whose values are and always will be unique. A *record key* thus has the ability to uniquely identify any given record within a file. In some files, it is desirable to have more than one record key, but normally only one key will be used in any processing operation.

In designing the files for a particular system, it is necessary to first understand the role of each file within the system. Three file classifications may be identified.

Master Files. A master file is a permanent file containing records essential for the day-to-day running of an organisation. Such a file must reflect the up-to-date state of an organisation's records and this is achieved by an operation known as updating which involves the processing of a master file against a transaction file.

Transaction Files. A transaction file is a file which contains records about the operations of an organisation within a specified time period. For example, withdrawal of money from bank current accounts are transactions about which information needs to be stored so that the pertinent customer records may be updated.

Backup Files. A backup file is a file which is required in case a master or transaction file becomes corrupt. This type of file does not actively participate in the data processing environment, but if corruption to a file does occur, by careful backtracking, the affected file can be reconstructed.

In addition to this broad classification of files, computer storage devices and programming languages draw a distinction between the physical sequence in which records are ordered on a device and the logical sequence of records which refers to the ascending values of the record key. Depending on the way that record keys are stored, there are two possible logical file orderings:

- sequential, in which records are stored in logical key sequence
- random, in which records are not stored in a logical sequence.

8.3.2 File Access and Process Modes

In deciding upon the various way in which records must be organised in a file, a system designer must consider design issues such as how are the records going to be *accessed* and how are the records going to be *processed*.

Accessing Modes

There are three possible ways for accessing a record within a file: by searching, using an index or using an algorithm.

Searching. Searching involves examining the records of a file until the required record is found. Several techniques are in common use and include linear and binary searching.

- The linear form of searching is a simple sequential search in which each record is successively examined, beginning at the start of the file, until the required record is found.

- The binary form of searching is applicable to sequentially ordered files and involves the successive division of the file until the required record is located. The record lying in the middle of the file is examined first to determine which half of the file contains the required record. The central record of the appropriate half of the file is then examined to determine which quarter of the file contains the required record, and so on until the required record is found. This process requires exact division, hence, if the number of records within a file is not 2^m-1 where m is any integer, then it will be necessary either to insert dummy records or to simulate such insertion.

Use of an Index. An index is a table which indicates the position of the record in a file corresponding to a given record key. A *simple index* consists of a table of record keys along with corresponding record addresses. In practice, a simple index is only used for small files. For larger files, *multi-level indices* in the form of tree structures are used. Thus, in the case of a two-level index, the first level index would define the range of key values relevant to each second level index and the second level index would define the record addresses corresponding to each key value.

Use of an Algorithm. In this case the record key is subjected to an algorithm which generates the corresponding record position within the file i.e.:

RECORD ADDRESS= f(RECORD KEY) where (f) is a discontinuous function

One problem that presents itself with this method is that of *synonyms*. There are two techniques available to a developer for dealing with synonyms.

- A synonym record can be stored in the next available record space in the main file area. This method is known as *consecutive spill*. A synonym record may be accessed by serially searching the file or by using pointers which chain each 'home record' to its corresponding synonyms.

- A synonym record can be stored in a special *file overflow area*. Retrieval of synonyms may again be via a serial search (of the overflow area) or by following a chain of pointers.

Synonyms are undesirable as their presence introduces an overhead when records are accessed and usually several algorithms may be tested so that the one which produces the least number of synonyms may be selected.

Processing Modes

A file processing mode is concerned with the order in which records are accessed during a series of accesses to the file during the execution of a computer program.

Those options available to a developer are as follows:

Serial Processing. In this mode, an entire file is processed in its physical sequence

Sequential Processing. Sequential processing is where an entire file is processed in its logical key sequence.

Selective-Serial Processing. With this processing mode, a subset of required records, known as hit records, are processed in their physical sequence

Selective-Sequential Processing. This form of processing is similar to selective-serial, except that the hit records are processed in their logical sequence

Random Processing. Random processing is substantially different from the other processing modes in so much as the hit records are processed in neither their physical nor logical sequence. In this mode, the precise ordering is determined by the application program.

8.3.3 File Organisation

The accessing and processing modes described in the previous section represent the processing requirements of the stored data. Based on these requirements a developer needs to identify and design file organisations which meet these requirements in an optimum way. Contemporary computer systems offer alternative storage organisations. Consequently, a developer must analyse the accessing and processing patterns required of a file and choose the most appropriate storage organisation. Much research work has been carried out in this problem area (*c.f. Hsiao & Harary, 1970*; *Cardenas, 1973*; *Severance, 1975*). The basic available options and the criteria used for their adoption are as follows:

Sequential File Organisation

When updating a master file which is sequentially organised the *brought forward/ carried forward* approach is followed. Transactions are sorted in the same order as the records in the master file. Firstly, a transaction record is read; records are then read in turn from the old master file. If a match for the transaction record is found then the master file record is updated otherwise no updating takes place. The next transaction record is then read and the process is repeated until the entire transactions file has been traversed.

Random File Organisation

With random file organisation, records are stored in no particular order and their positions in the file are indicated by applying an algorithm to the record keys. The suitability of an algorithm to a given application can be checked by software which applies the algorithm to the set of effective key values and counts the synonyms generated. Re-checking is needed if the set of effective key values changes appreciably. A randomising algorithm is designed with a view to creating an even spread of record addresses in order to minimise synonyms and gaps.

Random organisation has the advantage of enabling access to records directly, without having to access any preceding records in the file. Basic data operations such as *insertion*, *deletion* and *updating* can be handled efficiently in randomly organised files as these operations do not affect other records within the file. However, the problem of synonyms should be carefully considered during file design and their effects suitably minimised (*Hanson, 1977*).

Indexed Sequential File Organisation

Sequential file organisation provides efficient processing in those cases where the file concerned needs to be processed sequentially but performance suffers if the same file was to be accessed directly. In contrast, random organisation provides for fast direct accessing but this file organisation results in very poor performance for applications which require sequential accessing. A compromise solution is a method of file organisation known as indexed sequential organisation.

In an indexed sequential file, records are arranged in key sequence and additionally, their positions are indicated by an index. The index contains the key of a record together with its address. Records are stored within tracks on a disc and the index entry refers to a track number rather than an actual address. Within an index the key of the last record in the track is maintained. Therefore the mapping is of the form:

<number of track>:<highest record key in track>

In practice for most applications a two-level indexing system is used, comprising of a primary and secondary index. The primary index (cylinder index) is used to indicate the range of key values appropriate to each cylinder taken up by the file. For each such cylinder a secondary index (track index) indicates the range of key values appropriate to each track of the cylinder.

Inverted File Organisation

For the file organisation methods discussed above the information retrieval problem has always taken the form *find the records corresponding to the following key values*. In fact the techniques already mentioned are sufficient to cope with the demands of most data processing applications. However, in some cases the information retrieval problem takes the form *find the records which satisfy the following conditions on the values of the fields which make up the records*. For example, given a file containing information about students, a typical query might be as follows:

List the names and addresses of all students in their twenties who read Computer Science and take Management Science options.

One method of dealing with such a query would be to serially search the entire file, testing each record in turn against the given conditions. In general this is very inefficient and it is better effected by utilising an inverted file organisation. In this case appropriate indices are generated which indicate the record keys (or record addresses) corresponding to each possible value (or range of values) of the fields to which the retrieval conditions refer. For the example above indices for *age*, *course* and *option* would have to be created and maintained.

List Organisation

In some circumstances it is preferable or necessary to store a file as a linked list, where a list refers to an ordered set of records. The link between successive records is made via pointers included in the records. There are three basic types of list organisation available: simple list, double list and circular list structures.

Apart from the factors examined above, there are two further factors which relate to the performance of file processing and which need to be considered by a developer when choosing a particular file organisation (*Hanson, 1978*), the record *volatility* and the file *activity*.

Volatility refers to the number of records which are inserted to or deleted from a file during an application run. Highly volatile files will need to be reorganised frequently so that the available file storage can be used in an optimum way and the seek time is minimised.

Activity refers to the number of records which are updated during an application run. At design time knowledge about the activity of master files is only sketchy and therefore during testing and parallel running a developer will need to keep some statistics about the files and implement any changes if necessary. This may necessitate a whole restructuring of files and perhaps even choosing a different file organisation.

8.4 Characteristics of Database Schema Design

A database is an integrated collection of data records which are linked via logical relationships. In a database environment data is viewed as a whole, irrespective of its type. Data is not owned by a single department or application but is shared right across different functional areas.

At the heart of a database environment is a Data Base Management System (DBMS) which together with utilities such as data dictionaries, query languages and report generators provides the means of controlling the definition and manipulation of the data. Definition of data is achieved by the use of a Data Definition Language (DDL) whereas operations on the stored data such as insertions, deletions and updates are effected by the use of a Data Manipulation Language (DML). For a detailed exposition of the operation of databases and database management systems the reader is referred to (*Tsichritzis & Lochovsky, 1977*; *Ullman, 1980*; *Date, 1981*; *Oxborrow, 1986*).

The functionality of a DBMS is governed by an underlying data model which defines the allowable operations and imposed constraints upon the data structures. A major element therefore of the design process for a database is the specification of a schema and subschemas according to rules laid down by the available data model.

A schema comprises of data item types grouped into record types which are associated by relationships. Since in practice a schema is constructed for use in conjunction with a particular database management system, the choice of records and relationships must reflect the requirements of the software of that system. A schema can be described more precisely as follows.

A *schema* is a logical description of a database to be held in a computer and designed for a particular database management system. It gives both the data item types grouped into record types and the relationships that are to be stored. The schema is used by the database management software to specify the logical organisation and control of the database.

In most cases application programmers will wish to process only part of a database and require only a subset of the schema. This is called a *subschema* and it is used for limiting a particular application programmer's view of a database. Each programmer can manipulate the data only through a subschema and each application program includes a copy of a specific subschema in order to gain access to the database. More than one application program can use the same subschema and any subschema can be a view of the entire database. To emphasise the point, in a database system there exists only one schema but any number of subschemas, depending on the number of views required by the programmers.

8.4.1 Requirements from and Approach to a Schema Design

This section reviews the requirements form a schema design within a framework based on the ANSI/ SPARC architecture (*ANSI, 1975*).

The ANSI/ SPARC architecture considers databases from three points of view:

- the organisation as a whole (a single, organisation-wide view)
- the database user (application programmers and query language users)
- the physical representation and manipulation of data (the machine that stores the data and the database management system).

Each of the three viewpoints gives rise to different sets of requirements.

The first view requires a logical completeness of the data description, including any integrity constraints. This description must be sufficient to serve all potential applications within an organisation. It must also contain both semantics and structure of data.

The second view requires a data description which corresponds to the view of application programmers and users who will make use of query languages (note that such a view may be different to the first one). This description provides the user interfaces to the stored data.

The third view is concerned with efficiency of transaction processing and storage of data and for this a data description must be concerned with machine-level details.

These three views of database usage resulted in a three-level architecture:

- the conceptual schema
- the external schema and
- the internal schema.

The process of designing a schema can proceed following the above three level architecture.

The role of a conceptual schema is central to the design process and this is reflected in the framework adopted by this book. Conceptual data modelling, as discussed in chapter 5, will result in a conceptual schema which is a prerequisite for the design of internal and external schemas. An example of a conceptual schema from the hotel example was given in chapter 5 in figure 5.24.

Internal Schema

An internal schema would be designed to provide performance desired characteristics. The design activity associated with this has often been termed *physical database design*. Much work has been carried out in identifying the factors which influence performance.

There are two possible ways for *storing* records in a database.

Independently. A record is stored independently, when the DBMS uses the record's key to calculate a storage address within the allocated storage space.

In Association. The association between two records is used by the DBMS to store a record occurrence. This record is stored near its owner. The distance of this record's stored location from that of its owner depends on how many other member record occurrences have already been stored. If none, then the record is placed next to its owner.

Retrieval of database records may be accommodated in a number of different ways.

Using Indices. An index may be set up on primary keys for records stored by association or an index may be set up on secondary keys for records stored using either mode. The index table provides the mapping M:key→address.

Following the Storage Mechanism. If records are stored independently, they can be retrieved using a randomising algorithm on their key.

Alternatively, if the record association has been used, a member record may be retrieved by retrieving its owner and then searching through the stored set of member records until the required record is found.

Using Access Paths. Access paths are formed using the relationships between record types. A record may be retrieved using an access path irrespective of the mechanism used for storing it.

External Schema

An external schema serves as the means of defining the way users interface with the database through the DBMS. Each external schema provides an application view of the conceptual schema and therefore in general there should be many external schemas in a design specification. Although the ANSI/SPARC proposal advocates a general purpose mapping facility which transforms data between the conceptual level to external level, this proposal provides no details about these mappings. Consequently, no consensus view has emerged about the nature and structure of external-level schemas. Nevertheless, there are two factors which influence *navigation* around a database, entry points and access paths.

Entry points, are points in the schema structure at which an application program may begin its search for a particular record. The application program can then follow predetermined *access paths* from one record occurrence to another. Obtaining data for which a predetermined access path has not been established (from a given entry point) is likely to be very inefficient and may involve searching many records.

An access path may be formed in the following ways.

- By storing a member record near its owner record.
- By using pointers to link all related record occurrences . A physical pointer provides a physical address. A logical pointer is a value which can be used to obtain a physical address, either through calculation or look-up of an index.
- By using common values of data items. This method may be used when normalisation has taken place. A relationship between two entities is then shown through common values of duplicated data items in the two entities.

8.5 Logical Schema Design

In this section the hierarchical, network and relational data models are considered. In designing a schema, an analyst must consider the mechanism supporting the DBMS which is to be used by the organisation. Only then is it possible to cater for allowable operations as well as the imposed constraints which exist within the different database models.

It has been argued that the principal differences between database models are differences in the types of relationship permitted by any model and the restrictions imposed by each model on these relationships. The concept of *relationship* was covered in chapter 5. A developer in considering any database model should be aware of facilities provided by each model for relationships. For details on these differences the reader is referred to (*Clemons, 1985*).

8.5.1 The Hierarchical Model

The hierarchical model represents attribute relationships by considering record types and data items. Entity relationships are viewed as a parent-child relationship and the resultant structure is a restricted one in that certain entity relationships cannot be expressed. The unrepresented relationships can be established by constructing onc or more hierarchical definition trees which collectively will represent all entity relationships as defined in the entity-relationship diagram.

In every structure there exists one root record type at the highest level and any number of dependent record types at lower levels. A parent is a record occurrence that has subordinate record occurrences called children. Each record occurrence, except for that of the root record, must be connected to an occurrence of a parent record.

Hierarchical DMLs can be classified according to the way they operate on the hierarchical definition tree. There are two popular methods currently employed, and these are tree traversal and hierarchical selection.

Tree Traversal. Tree traversal, is a method by which a tree structure is searched in a specified order. This method, chooses record occurrences from *top to bottom, left to right.* This order can be physically represented either by a sequential or by a direct data organisation.

Hierarchical Selection. Hierarchical selection is a method by which records are selected according to some relationships between data items.

As an example of a hierarchical data schema, consider figure 8.11. The diagram represents a hierarchical data schema for a subset of the hotel conceptual schema. This subset of the conceptual schema has been mapped onto a structure in which the lines are directed, pointing from 1 to N in the 1:N relationship and is additionally restricted by representing an ordered tree.

The mapping from conceptual schema to hierarchical schema is neither unique nor algorithmic. A designer needs to consider the nature and type of transactions, knowledge acquired from the process design model, and based on this knowledge determine the placings of the entities (records) in the tree structure. For example, frequently accessed records will be placed at or near the root node since access of root record types is usually faster and more efficient than access of records which are dependent on other records.

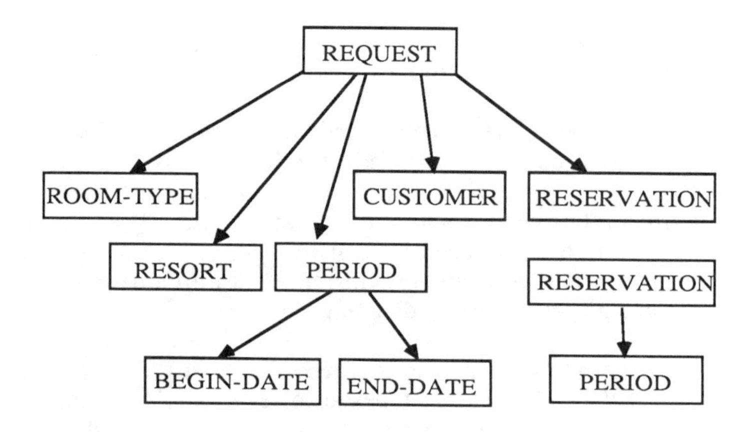

Figure 8.11: A Hierarchical Schema

8.5.2 The Network Model

A network database model, as the name implies, is represented as a network where the nodes correspond to entity types and the arcs correspond to relationship types. The network model has been comprehensively specified by the Data Base Task Group (DBTG) (*CODASYL, 1971*) and the original specification has received numerous revisions (*CODASYL, 1973, 1976, 1977, 1978*).

The basic structure within the network model is the *DBTG set type*. In linking two record types, a hierarchy is established between these record types by naming one as the *owner* and the other as a *member*. This basic structure between record types is called a set type. Using set types, the database designer can design a complete schema.

Set types provide logical links between record types and define the paths along which the system can travel in order to locate the data. The following points must be considered when set types are constructed.

- A record type may be an owner in one set type and a member in another set type.
- A record type may be a member in more than one set type.
- There may be more than one set type established between two record types.

Considering again the subset of the hotel conceptual schema. A network schema will be represented as shown in figure 8.12.

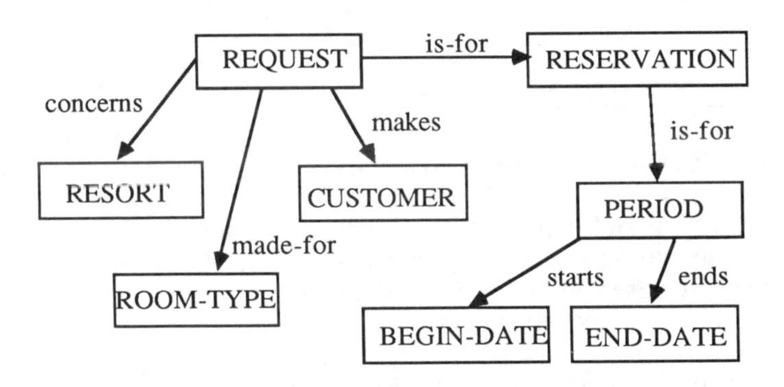

Figure 8.12: A Network Schema

Knowledge of transaction behaviour is necessary to complete the design of a network schema. This knowledge affects the navigation through the data structure and determines the way that records enter the database and how they are manipulated.

When developers designs a network schema, they must make decisions on how member record occurrences will enter the database (*insertion mode*) and later, how they can be manipulated (*retention mode*). For the insertion mode there are two choices which are: *automatic* and *manual.*

When automatic membership is chosen for a set, each occurrence of the member record type is connected to that set automatically. Manual membership is chosen when each occurrence of the member record type is not automatically connected to the set but the connection must be made by the application program explicitly via a DML command.

For the retention mode there are also two options, which are *mandatory* and *optional*. Mandatory membership means that occurrences of a record type cannot be removed from any occurrence of a set type once they have been connected. Mandatory record occurrences can be moved by DML statements, from one set occurrence to another but it must be within the same set type. Optional membership is chosen if record occurrences are allowed to be removed from set occurrences by DML commands.

With regard to physical storage considerations the analyst has the choice- in the schema definition- as to how record occurrences are to be organised and thus can obtain a variety of ways in retrieving record occurrences. The most popular of these methods are the use of an attribute (or set of attributes) to calculate a record address, known as the CALC method, and the use of the logical relationship between two records known as the VIA SET method.

8.5.3 The Relational Model

The major difference between the relational model and those of hierarchical and network, is the structural way in which relationships are expressed. Both hierarchical and network models use record types to represent entity types and links to represent entity relationships. In the relational model an entity type is represented as a relation. A relation can be thought of as a table closely resembling a sequential file.

The existence of a relationship between two relations depends on the same value of a data item being in each relation.

There is no standard way for specifying relationships in a schema. Also physical storage descriptions may vary. For example, a relation may be regarded as a file with the possible use of indices on secondary keys to express relationships between relations.

Relations may look like ordinary sequential files but where they differ dramatically from files is in the way that these relations can be operated upon. Relational operations, unlike file operations, operate on entire relations rather than on one record at a time.

8.6 Summary

There are many ways in which data may be stored. The design of data needs to be considered from both logical and physical perspectives.

The logical design of records is a necessary activity so that there are no side effects when the basic data operations of insertion, deletion and accessing operate upon a set of stored data. This chapter has demonstrated two possible ways of carrying out such a design: data normalisation and direct mapping from a conceptual schema.

The influence of the physical characteristics of files on the manipulation of data is a direct result of the multiplicity of different storage organisation and accessing modes. Each file organisation has certain advantages over the others, relative to the way the file is to be used within the computer system. This chapter has examined the advantages and disadvantages of the way data may be organised within the context of file and database structures.

Chapter 9

Interface Design

In the previous two chapters, attention was paid to the design of the underlying data structures within an application and to the design of the software modules that will manipulate that data. In this chapter, attention is turned to the interface between a software system and its users.

The area of interface design is normally of crucial importance in the design of a system because, for most system users, this is the only aspect of a system that they will be aware of. Consequently, many systems have in the past been rejected not because of poor performance or their inability to meet a user's requirements but because of a poor interface which leaves users confused and frustrated.

This chapter begins by examining the principles of good interface design and after reviewing the nature of interface devices, looks at the notion of interactive dialogue design. Typically, a large proportion of an information system will be concerned with interaction between the system and its users and it is therefore necessary to develop a coherent model of this user interaction prior to system implementation. Finally, the chapter looks at the nature of interfaces in terms of the data that passes across them, with special reference to the validation of this data and the manner in which it is coded and presented.

9.1 Principles of Good Interface Design

Before considering particular aspects of interface design, it is important to first establish the principles of good interface design. The following act as a set of general design guidelines.

Establish a Clear User Model. All users will perceive a system as a model, such as a payroll system, order processing system etc. Because the system interfaces (screens, reports etc.) are all that a user sees of a system, it is important that the interfaces help a user to develop a clear model of the functioning of the system. In the initial stages of systems analysis, the analyst will have developed a *task analysis* model of the user's view of the system and this can therefore play an important role in the development of a system interface. Other important considerations for system interfaces are as follows.

- Irrelevant detail such as internal screen numbers or report numbers appearing on output should be avoided. Generally these serve no useful purpose to users and simply confuse and clutter the output.

- Adopt a consistent approach to the use of screens and terminal keyboards. Typical examples of standard function key uses are the ESCAPE key to permit a user to abort the function currently in operation and function key 1 to provide on-help.

- Communicate with the user in a language with which they are familiar, avoiding computing jargon. When a file or record is unavailable for processing, relate the problem to the task the user is currently trying to perform, rather than an abstract message such as "Cannot read record". In particular, avoid the use of meaningless codes and input values.

User Guidance. At no point in a system should the unfamiliar user be left without some indication of how to proceed. Typical considerations should include the following.

- Clearly state the stage of the system the user has currently reached and what alternatives are available, possibly by displaying a menu of the alternatives.

- Indicate what action a user is expected to take at any point in the system by either displaying a prompt message such as "Select an option" or employ a standard set of defaults, such as one where the user must select a function key.

- Provide of a help facility to assist novice users. The help provided should always be context-sensitive; that is the help information should relate directly to the function or choice of functions currently available to the user. General help information which does not change as the user selects functions can be unhelpful and confusing.

User Reassurance. During processing or on receipt of a user command, a system should inform a user of the action it is taking or is about to take. Thus systems should attempt to support the following.

- All input routines should acknowledge all input, either by displaying a message stating such or ringing the terminal bell to indicate successful input, as with electronic cash tills found in most large shops.

- Systems should display the data currently being processed, such as before the deletion of a record. Furthermore, in the case of a deletion, a system should ensure that the current record is indeed the record the user wishes to delete.

- Functions should indicate the action currently being undertaken by the system, especially when the operation may take some time. This can best be achieved by displaying a simple message, such as "Working..." during the processing of data.

Error Messages. Error messages should always make it absolutely clear what error condition has arisen, why it was caused and what action should subsequently be taken. In particular, designers should ensure the following error handling characteristics.

- Programs should distinguish between errors resulting from the user and which are within the user's ability to correct, such as entering an invalid code number, and errors which have arisen from system constraints, such as data tables which are too small, in which case the user should be informed of whom to contact

- Distinct error messages should be assigned for different errors; one technique is to assign each possible error with a unique number and the name of the program component which generated the error; compound error messages should be avoided

- Full explanations of an error and any corrective action necessary should always be given.

Having established these basic interface design principles, the remaining sections of this chapter concentrates on particular techniques and approaches to their use.

9.2 Interface Devices

Any discussion of interface design must include some consideration of the devices which will form the interface between machine and user. The importance of this topic arises because data capture has long been recognised as the weak link in the chain when introducing computerised systems, since the ability of the human operator to enter data into a system is far outstripped by the processing power of most modern day computers.

With the advent of the commercial computer system, the expense of the equipment meant that new systems tended to be remote from their users. Typically, systems such as stock control and payroll required the completion of handwritten documents which were subsequently passed to data preparation clerks who transcribed them into machine readable formats. It was this transcription process which proved for many installations the bottleneck in the systems and devalued the benefits of computer-based systems. With the increased availability of relatively cheap computer hardware it has been possible to remove the labour-intensive and unnecessary step of data preparation by allowing users to enter data to computer systems, at source. For example, in the hotel case study, front-desk clerks can enter data directly into on-line terminals as and when guests arrive or depart.

Of course, computer terminals are not the only type of data entry device, and the following sections review the different devices available.

9.2.1 Punched Media

Punched cards have been used as an information recording medium for over 100 years, but today, as with paper tape, they are largely obsolete. The advantage of cards and tape was that they were capable of off-line preparation, thus freeing a computer for more processing work. Furthermore, punched cards have the advantage of being conceptually straightforward, readable by humans as well as machines and can act as source documents. Unfortunately, the machines needed to prepare and read such documents are expensive, are often unreliable and frequently damage the cards or paper tape. This, combined with the advent of more efficient data capture devices and the need for large data preparation teams led to the demise of punched cards and paper tape.

A variation on the punched card/ paper tape theme, which is still widely used is the kimball tag. The kimball tag is a small cardboard tag containing small perforations similar to paper tape. Normally kimball tags are attached to shop goods and the perforations represent the good's stock number. Thus when an item is sold, the tag is removed and placed in a special reader which determines the stock number. This information can subsequently be used to replace the goods in the shop in the context of a computerised stock control system.

9.2.2 On-Line Computer Terminal

The advent of cheaper computer terminals, the most familiar being the visual display unit (VDU) led to a complete change in the philosophy of data capture- no longer must data be prepared off-line, transcribed from handwritten documents but instead could be entered directly through a computer terminal into the computer system.

The advantage of the VDU is that it is a two-way communication medium, that is it permits the computer to communicate with a user, as well as the user communicating with the computer, thus enabling a system to prompt and assist users during data input.

VDUs consist of a standard *qwerty* keyboard and a monochrome or colour screen, normally 80 characters by 24 lines. More advanced VDUs permit screen attribute painting in which software can request certain areas of the screen to be highlighted, dimmed or flash, thus permitting the development of advanced interfaces. In addition, keyboards are often extended to contain extra function keys, whose use can be detected by software and treated in some predetermined way.

9.2.3 Workstations

The disadvantage with computer terminals linked directly to a computer system is that part of the computer's processing power is tied up with the relatively simple operation of data input and output. In order to free computer systems from this task and to allow their full dedication to processing, the workstation concept has evolved.

Figure 9.1 shows a system diagram of a workstation approach to data entry. At the centre of the workstation system is a dedicated micro-processor whose function it is to perform simple data input, data storage and printing operations. Connected to the micro-processor are a series of workstations, which are often only VDUs. For the majority of the time, the workstation system can accept data through the individual workstations and after preliminary validation to check for completeness and consistency, the data is stored. At certain times, often overnight, the micro-processor can establish a link to the main computer system and subsequently pass all the stored data to the computer for processing.

The advantage of this approach to data input is that the collection activity is performed independently of the main computer thus freeing resources on the computer and, should the main computer fail, permit data input to continue.

The workstation approach is commonly adopted in building society and banking systems in which a conventional direct terminal approach would require literally thousands of terminals to be connected to the computer causing severe communications processing overheads.

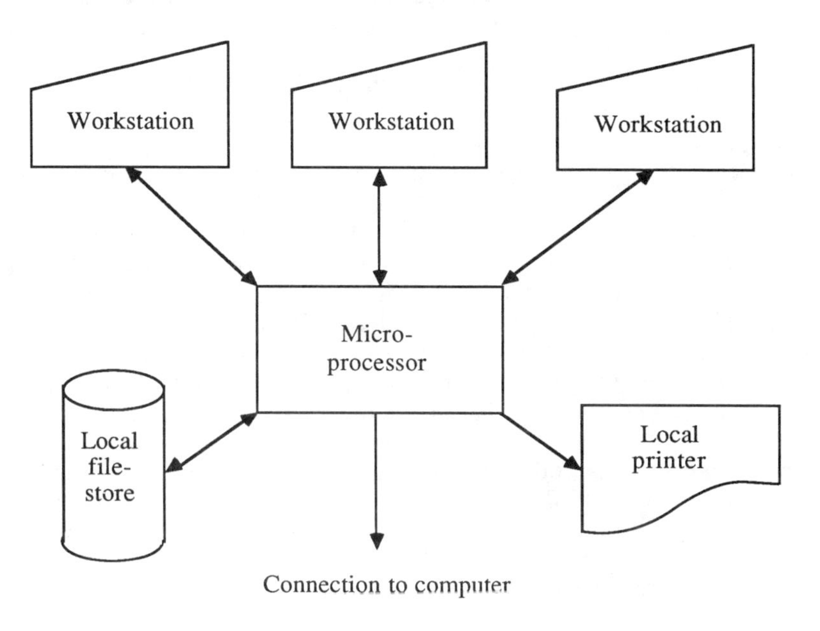

Figure 9.1: Data Entry Workstations

9.2.4 Pre-Printed Character Recognition

In many applications, the range of data for input to a computer system is limited. For example, a shop may number all its stock lines and when an item is sold, the stock number is entered into a computer in order to generate a purchase order to replace the item of stock. In such applications, where the choice of data values is limited, the use of pre-coded tickets or pre-printed forms is appropriate.

A variety of devices and techniques exist for the use of pre-printed character recognition techniques. The following are the most commonly available.

Magnetic Ink Character Recognition. The technique of magnetic ink character recognition (MICR) is based upon stationary which has been pre-printed with magnetic ink in one of several alphanumeric character fonts. This has the advantage of being both readable by humans and by machines, the latter employing special readers. The most common application of MICR are bank cheques which are pre-printed with cheque number, branch sorting code and account number. The value of the cheque is normally added when a cheque is presented for clearing, using a special machine known as an inscriber. Cheques are subsequently placed in special readers for processing.

The disadvantage of this technique is that printing must be of high quality, using special ink and precise positioning on input documents.

Optical Character Recognition. Optical character recognition (OCR) is a more general case of MICR in which character recognition is based upon characters pre-printed in standard ink, but from a limited set of fonts. Whilst not as widely used as initially predicted, OCR has a well established use in high-volume applications. For example, gas and electricity bills can be issued, pre-printed with the consumer's details and value of the bill. On receipt of payment, the bill can be input to the accounting system for recording as being paid. Such documents are usually known as turnaround documents.

Bar Coding. Bar coding is now one of the most popular methods of rapid data entry and is widely used in product labelling for stock reordering and point of sale systems. A bar code, such as that shown in figure 9.2, simply consists of a number of vertical lines or varying thickness, which represent a number value.

Figure 9.2: A Sample Bar Code

For convenience, the numeric value is generally printed beneath the coding. The bar code value is read into a computer system through the use of a wand or laser scanner which transcribes the bar coding into the appropriate value.

9.2.5 Menus, Mice and Icons

A common data input mechanism for limited range data is the use of menus and lists, typically for command input to a system. A menu is simply a list of all possible alternatives, or categories of alternatives in the case of a large number, which are referenced by the user. Referencing techniques may vary from each alternative being given a code which is entered on a keyboard by the user, to the use of light pens or screen pointing to identify the chosen option. A cheaper alternative to the light pen is a mouse, a small device which when moved around on a flat surface, correspondingly moves a cursor around a VDU screen. When the cursor rests upon the required menu alternative, a function key on the mouse is pressed to indicate the chosen alternative.

An increasingly popular menuing system is the use of *pull-down menus*, in which a summary menu appears, usually across the top of a screen. When the user points to one of the options with a mouse and selects it, a sub-menu appears below the summary. The user can then point to an item on the sub-menu and select it to perform the desired function. An example is shown in figure 9.3.

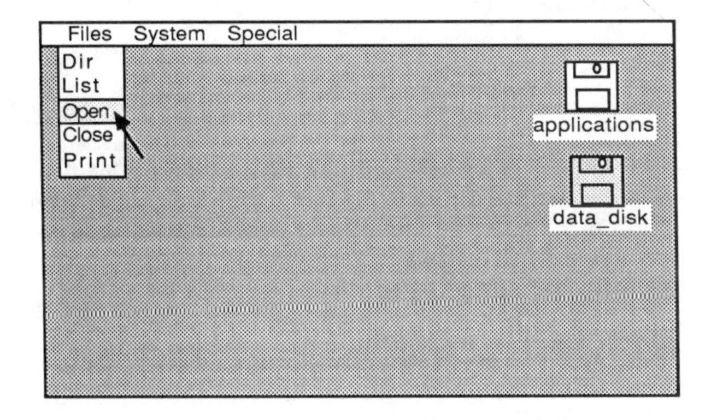

Figure 9.3: A Example Screen with Menus, Pointing and Icons

A final method of command selection is through the use of icons. An icon is a symbolic image used to represent some object within a system. In figure 9.3, two icons are shown on the right-hand side of the sample screen, to represent the presence of sub-directories in a disk filing system. The names of these directories are *applications* and *data_disk*. Such icons, when selected, may cause the contents of each sub-directory to be listed or if the icon represented an application program, for that program to be invoked.

9.2.6 Speech

Speech recognition and speech generation are still in the early stages of development. However a number of techniques exist for limited speech recognition and several research and development projects are underway to improve and widen the scope of these techniques.

9.3 Interactive Dialogue Design

The design of any interactive, on-line system will always necessitate the production of a dialogue design model. The aim of such models are to ensure that a clear, logical user model exists within a system and thus meet the first dialogue design requirement outlined above.

Most early definitions of structured development methods paid little attention to explicit dialogue design, leaving it to 'drop out' of process design models. However, with greater awareness of the need for proper dialogue design, several methods are beginning to employ a variety of techniques for dialogue modelling (*Longworth & Nicholls, 1986; LBMS, 1987*). As with process modelling, an analysis of the various dialogue modelling techniques reveals an underlying similarity in terms of the concepts they use and it is these basic concepts which are now described.

9.3.1 Basic Concepts

At the heart of an interactive dialogue lie a number of basic concepts, which together can be assembled to form complex system-user dialogues. The most basic of these concepts are the menu, user decision and action.

Menu. Menus are formatted text and data displays which indicate the possible processing options available to the user at the current point within the application. Typically, alongside each menu item will be an indication as to how to invoke the item, either by pointing to an icon, pressing a function or typing a command. Menus are shown as a six-sided box in dialogue models.

User Decision. A user decision is simply the point at which a user selects a processing option by whichever input mechanism is provided. Its result will be the invocation of a control path leading to another menu or the firing of an action. User decisions are shown as triangles in dialogue models.

Action. An action corresponds to some discrete system behaviour, the effect of which may change data values held by the system or to pass control to another dialogue structure. A number of system-defined actions may be used, such as EXIT (end the dialogue) or RETURN (go back to a previous menu structure). Actions are shown as ovals or rounded squares in dialogue models. Ovals are generally reserved for system-defined actions.

In addition to these three basic primitives, there is the concept of a control path, which links the primitives together to form a dialogue structure. A control path may link user decisions to processes or menus; link menus to user decisions and link processes to menus.

9.3.2 Dialogue Models

The basic dialogue modelling primitives can be assembled to form graphical representations of dialogues. Figure 9.4 shows an example of two dialogues, which are related to each other.

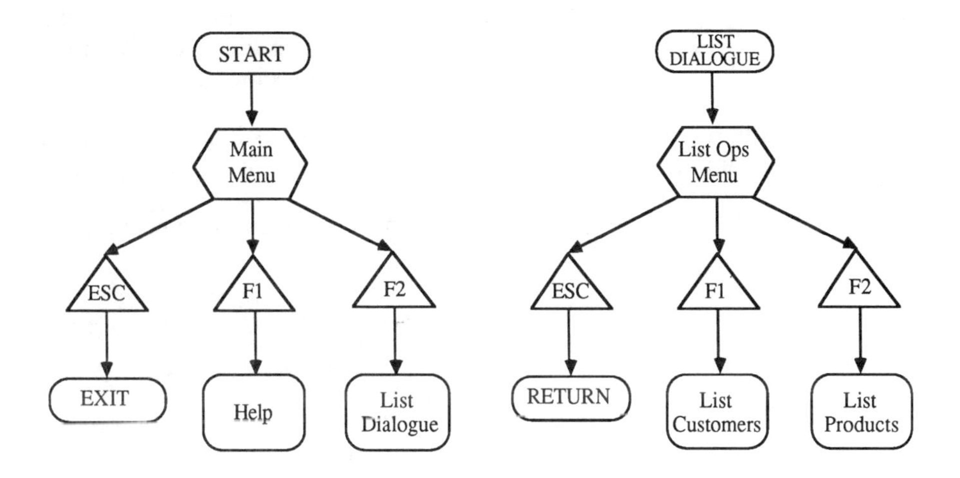

Figure 9.4: A Simple Dialogue Model

The first dialogue has the name START and the second is called LIST DIALOGUE. START is a special dialogue name, indicating that an application program begins with this dialogue, whilst LIST MENU must explicitly be called from some other dialogue structure.

The START dialogue shows that when invoked, a menu called MAIN MENU is displayed. The user has the option of three decision paths, selecting the ESC key, the F1 key or the F2 key. In the case of selecting the ESC key, dialogue control passes to the system-defined action, EXIT, which causes the dialogue to terminate.

In the case of a user selecting the F1 key, control passes to an action called HELP. As there is no dialogue structure called HELP, it is implied that HELP is a process. After the execution of the HELP process, control returns to MAIN MENU.

In the case of a user selecting the F2 key, control passes to an action called LIST DIALOGUE and, as a dialogue structure with the same name exists, control is passed to that structure.

The LIST DIALOGUE structure shows all the same concepts as the starting structure, except it contains the RETURN action, the effect of which is to return control of the dialogue the menu of calling action, that is MAIN MENU.

In some dialogues it may be desirable to attach conditions to the execution of a particular control path. Conditions can be shown by altering the user decision symbol to a conditional decision symbol, containing not only the user decision, but also any preconditions which must be satisfied. Failure to satisfy the conditions would result in the user decision being rejected.

The example in figure 9.5 shows the user of the conditional decision symbol. The action resulting from the selection of the F2 key is dependent upon whether today is currently Friday or not. Notice that the conditional decisions are mutually exhaustive; however, this does not need to be the case.

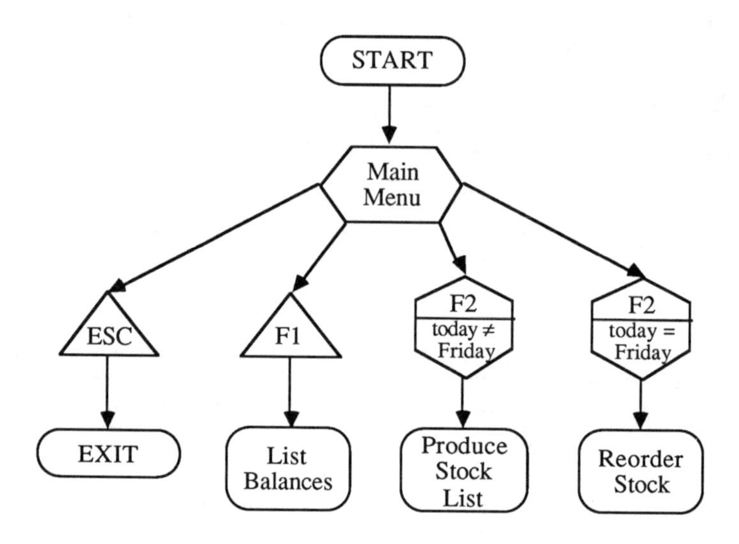

Figure 9.5: Conditional Dialogues

9.4 Data Classification and Coding

One of the problems faced in any data processing system is the gap that exists between how a user views data and how a computer program views the same data. On the one hand, users typically view data in arbitrarily sized lumps and often referring to it by lengthy, English, names such as "a 2 inch, rounded headed screw". On the other hand, computer programs are more easily written when handling fixed-sized packets of data that are well structured and having a unique name. Thus the round headed screw referred to above, may be assigned the stock code number 20301.

Unfortunately, neither human nor computer representations are particularly suitable for the other to use. The English description, whilst totally understandable to the human, would be extremely difficult for a computer to process. Whilst the reference 20301 is a compact, precise code for a stock item, but it is not possible for a human reader to easily understand what stock item the code number references. Thus during data input, a user may easily make a mistake without realising it.

The problem that system designers face in data classification and coding is in choosing a scheme which both permits the use of precise codes for computer use and yet which is understandable to the human. Before examining the different methods of code design, it is useful to first further identify the requirements of a good code.

9.4.1 Code Requirements

A good code should meet the following requirements.

- An entity must be capable of having a unique code number in order to avoid the possibility of misidentification
- Code numbers should be as brief as possible, subject to readability and desired redundancy
- Codes should be constructed in such a way as to minimise errors when used, thus consecutive code numbers such 3576 and 3577 should be avoided since mistyping may cause the trailing 6 to appear as a 7; likewise codes containing similar characters such as 85E and 85F should be avoided, since E may easily be mistaken for F
- Long codes should be broken down by delimiters such as comma, slash, hyphen or period
- Code systems should be capable of expansion.

9.4.2 Coding Systems

A variety of coding systems exist, the most common of which are described below.

Serial Coding

Serial coding schemes employ no technique in the allocation of code numbers to entities, with the exception of avoiding using one of the code numbers in pairs such as 123 and 132, where the 2 and 3 could easily be transposed or 85E and 85F where the E could be misread as an F. The advantage of these codes is that they are simple and relatively efficient since they have little redundancy. However, since the code numbers convey no information about the characteristics of the entity which they encode, serial codes are not widely used.

Range Coding

Range or block coding represents a minimal attempt to classify entities into groups which have some common attribute. Typically a range of code numbers will be set aside for each group of entities. Code numbers within each group are then allocated on a sequential or serial basis. For example, hotel room numbers are often allocated so that the first digit represents the floor on which the room may be found:

101-109	:	rooms on floor 1
201-215	:	rooms on floor 2
301-315	:	rooms on floor 3
		etc.

The main problem with range codes is ensuring that enough code numbers are left for the addition of new entities.

Hierarchical Coding

Hierarchical coding takes the concept of range coding a stage further. The first part of a hierarchical code is used to broadly classify entities into a number of groups. The second part of a hierarchical code is used to subdivide each group into categories, followed by the third part of the code which subdivides the categories and so on.

For example, hotel accommodation types may be classified using the scheme shown in figure 9.6. Using this scheme, an economy, apartment with no bath is assigned the code *132*, in which *1* indicates that the accommodation is an *apartment*, *3* indicate that the accommodation is *economy class* and *2* indicates that there is *no bath*. Note that the code is interpreted from left to right and that the interpretation of each code digit is dependent upon the interpretation of the previous digits. Thus in the code number *122*, the last 2 is interpreted as a *no separate room* and not as *no bath*.

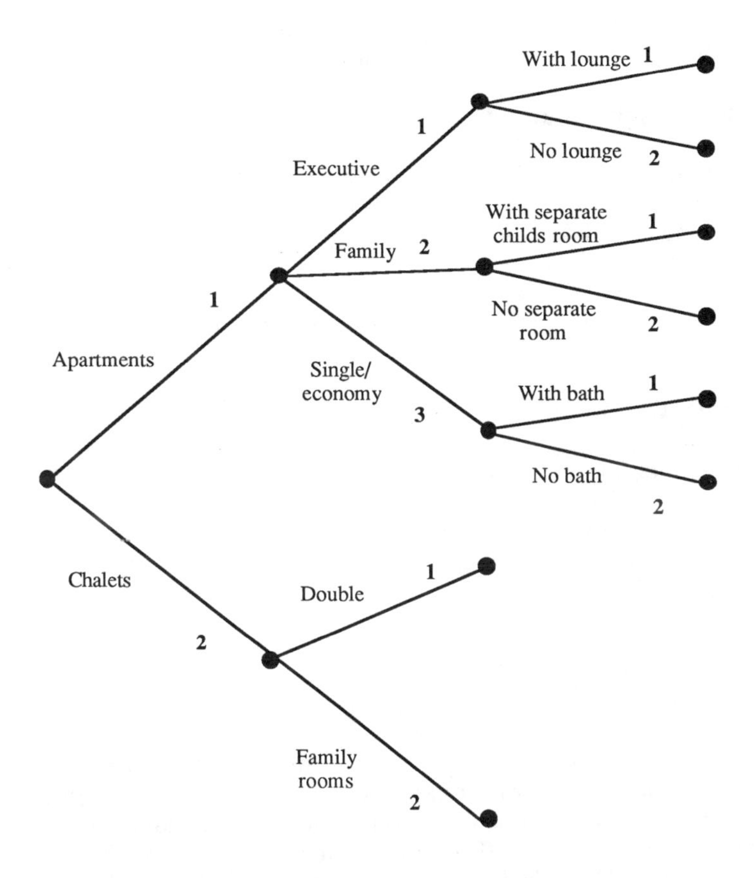

Figure 9.6: A Hierarchical Coding Scheme

Faceted Coding

Faceted codes are similar to hierarchical codes, except that the interpretation of each code character is not dependent upon the interpretation of previous parts of the code. Figure 9.7 shows a faceted code for the hotel accommodation classification example. Using this code, the code number for an economy apartment with no bath *136*.

First digit (accommodation type)

1= apartments	2= chalets

Second digit (size)

1= executive	3= single/ economy
2= family	4= double

Third digit (lounge)

1= with lounge	4=no separate childs room
2= no lounge	5=with bath
3= separate childs room	6=no bath

Figure 9.7: A Faceted Coding Scheme

Mnemonic Coding

Mnemonic codes are codes which use an abbreviation of an entity's name as the code value. For example, a resort code for hotel location in London may be LON and MAN for Manchester. The advantage of this type of code is that through careful choice of mnemonics, code values can be suggestive allowing easy coding and decoding by humans.

9.5 Data Validation

Another consideration for the system designer when designing system interfaces is the problem of erroneous data and an important element of all information system design is the incorporation of data validation techniques into the system design. The importance of this task should not be under-estimated since if erroneous data is input to a system, the system will at best either fail or produce meaningless results and at worst, will damage or corrupt files and databases used by the system.

Several checks exist for the validation of data and these are outlined in the following sections.

9.5.1 Type Checking

Type checking is concerned with ensuring that data belongs to a particular class. For example, car registration plates often consist of 3 letters, 3 digits and a further single letter. In a data validation system, the possible valid set of car registrations would be too large to enumerate explicitly, so instead a type check based on the above characteristics could be performed.

9.5.2 Range Checking

An extension of the type check is the range check. In type checking, only the class of data is considered, whereas in the range check the value of the input data is also considered. For example, percentage examination marks would normally be in the range 0 to 100, which implies not only a type check, i.e. that the mark is numeric, but also limits the range of permitted values.

9.5.3 Check Digits

A check digit is a digit appended to a code number which is used to detect errors which arise when the code number is incorrectly entered into a system. Several methods of check digit protection exist, the most common being the modulus-11 check digit system, which is used on international standard book numbers which are found on every published book.

The modulus-11 check digit system gives a high level of security and can usually detect the majority of errors which arise through the input of an invalid code number.

The method requires that once code numbers have been selected, a check digit must be appended. This is achieved using the following algorithm.

1. Give each digit in the code number a weight starting by assigning 2 to the least significant digit and increasing the weight by 1 for each next significant digit.

2. Multiply each digit by its weight and sum the products.

3. Divide the sum of the products by 11 and subtract the remainder from 11 to give the check digit. If the remainder is 0, the check digit is also 0 and if the remainder is 10, the check digit can be 0 or the character X, according to which ever convention you wish to adopt.

Figure 9.8 shows an example of the generation of a modulus 11 check digit for the code number 27935.

Code	2	7	9	3	5
Weight	6	5	4	3	2
Code * weight	12 +	35 +	36 +	9 +	10

= 102

102 divided by 11 = 9 remainder 3

Check digit is given by 11 - 3 = 8

New code number is 2 7 9 3 5 8

Figure 9.8: Calculation of a Check Digit

When validating a code number, a similar algorithm is followed.

1. Assign weights to each digit in the code number, including the check digit which is given a weight of 1.

2. Repeat step 2 above.

3. Divide the sum of the products by 11 and subtract the remainder from 11. If the remainder is 0, the code number is valid, otherwise it is invalid.

9.5.4 Compatibility Checking

The methods of data checking outlined so far examine data values in isolation. Sometimes however, two items of data may have a relationship and thus the data can be validated in the context of each other. For example, a stock control system may record the cost price and sale price of stock items and assuming that the organisation running the system always sells a stock item at a price greater than its cost, a simple compatibility check of ensuring that sale price is greater than cost price can be implemented.

9.6 Data Output

So far, the techniques of data classification, coding and validation, together with the supporting hardware devices have been considered. Attention is finally turned to the output of data from a system and in particular how this data should be formatted.

9.6.1 Screen Design

Screen design is concerned with the formatting, layout and presentation of information on a VDU screen. The initial step in screen design is to perform an analysis of the dialogue that will take place between the user and the machine. This analysis will require the identification of conversational blocks- logical groups of dialogue that are concerned with a single activity, such as the input of a customer order or the displaying of details about a stock item.

For each conversational block the following guidelines should be employed.

Logical Sequencing. All information relating to a single, atomic function performed by the system, should appear on the same screen.

Relevance. Only information that is relevant to the current operation of the system should appear on the screen.

Consistency. A consistent style of layout and language should be used in order to minimise the need for users to stop and think about what the system is saying to them.

Simplicity. Displays should be kept simple and screens should not become overcrowded. Generally only 40% of the screen should be in use.

Coding. Where some information is more important than others, its display should be coded through the use of highlighting, colour, reverse video or flashing. When using advanced VDUs the use of graphics and picture images can improve communication efficiency.

9.6.2 Report Design

A wide variety of types of report can be generated by a computer system, each of which will require different design and layout characteristics, although all should adhere to the general principles of good interface design previously described, together with the following general principles of good report design.

- Align headings above the centre of a field.

- Justify fields- numeric fields should be right justified and aligned on the decimal point whilst alphanumeric fields should generally be left justified.

- Highlight important fields using a bold type face, capitals or underlining.

- Optional fields should be spaced filled when unused, but large areas of blank space should be avoided.

- Number pages for reference and ensure that pages are not destroyed during a print run; date and time stamping is also recommended.

An example of a report following these guidelines is shown in figure 9.9.

```
HOTEL ADMINISTRATION SYSTEM                  9 JUL 88   10:13    Page  1

                           Guest Account Transactions

Room  Guest's name    Arrival  Credit        Transaction Details
                               Limit       Date    Type          Amount

101   Jones           7.7.88   2000.00   7.7.88    B/f             0.00
                                         7.7.88    Room           52.00
                                         8.7.88    Breakfast       6.50
                                         8.7.88    Room           52.00
                                         8.7.88    Paid          162.50 -

                                                   Total           0.00
103     Wilson        8.7.88   2000.00   8.7.88    B/f           133.23
                                         8.7.88    Restaurant     47.38
                                         8.7.88    Bar             7.52

                                         continued
```

```
HOTEL ADMINISTRATION SYSTEM                  9 JUL 88   10:13    Page  2

                           Guest Account Transactions

Room  Guest's name    Arrival  Credit        Transaction Details
                               Limit       Date    Type          Amount

103     Wilson        8.7.88   2000.00   9.7.88    Room           52.00
                                         9.7.88    Restaurant     13.92

                                                   Total         244.05
```

Figure 9.9: A Sample Guest Transactions Report

Various types of report can be generated by a system and these can be summarised as:

- listings reports, which are simple iterations of a record

- transaction reports, which contains several groups of related data items and supporting information

- block structured reports, which are used for information reporting which have several different formats but preserve some block structure of related items.

The first step in report design is to establish the type of report. Because interfaces are one of the few aspects seen by future users of a system, report design is one of the design activities in which users can actively participate and it is at this stage that a rough report layout can be sketched, in collaboration with users.

The next step is to apply detailed design guidelines to the report layout, depending upon the type of report.

Listing Reports

Listing reports are the simplest kind of report. Data is normally ordered in a column format, the key field appearing on the left hand side. Subsequent data fields should appear in order of importance, as determined by the user.

Because listing reports tend to be lengthy, design decisions must be made concerning the pagination of the report. Long reports should always be paginated with a heading and trailing space and pages numbering. In some instances, it will be useful to produce sub-totals at the foot of each page.

Transaction Reports

Transaction reports usually have a form type of layout which consists of several logical blocks of information. An example of a report of this type is shown in figure 9.10, which is a listing of a hotel guest transactions file taken from chapter 7.. The ordering of blocks will be determined by the logical sequence and their relative importance. Most transaction reports have a heading block followed by the report body and a footing block. Thus in the above example, the heading block information is room number, guest's name, arrival date and credit limit, whilst the footing block is a total of the value of charges. Repeating items in the report body may require a page break, if so, it is important to label the continuation copy by repeating the header information. Blocks should be spaced in an order which helps reading of the report.

Block Structured Reports

Block structured reports are generally the most complex to design because of their varying nature. Under these circumstances the general principles outlined above should be employed.

9.7 Summary

Interface design is one of the most critical parts of a system in so much as it represents the only part of a system that a user physically sees. Therefore, the subject merits careful attention from the system designer. A summary of those items that require consideration are:

- the type of interface device to be used
- the design of the system-user dialogue in an interactive system
- the nature of the data and the most appropriate way to code it
- the data validation techniques that will be required
- screen and report layout.

Chapter 10

Planning and Control

In this chapter, attention is turned to a number of issues which can be regarded as less technical than those previously addressed and yet are vital to the successful development of any large system. The issues involved fall broadly into two categories. The first concerns the issues involved in moving from a coded system through to the system running in a *live*, user environment. Whilst the second category examines the issues of project management which must underlie the entire development.

10.1 System Testing

Before the introduction of a new system, developers will need to devise a strategy for ensuring that a system meets the objectives and functionality laid down in its requirements specification. There is a natural belief that the activity of testing belongs to the task of programming task. However, experience consistently shows that products, including software, need to be tested, not only by their immediate producers, but also by those who specify their requirements.

In specifying test plans, the system developer needs to consider three levels of testing.

Logic Testing. This form of testing can be regarded as testing at the lowest level of detail- do individual system modules perform the required function? This testing is usually carried out by programmers.

Program Testing. This level of testing is concerned with ensuring that modules assembled into programs operate in a coordinated and correct fashion. In particular, tests are performed to ensure that modules receive the correct control and data input parameter values and that they generate the correct output parameter values.

System Testing. Once all the modules and programs within a system have been tested, they can be assembled into a complete system, which in turn requires testing. System testing also involves testing the complete computer system, hardware and software, as well as the supporting manual activities and documentation.

Within any of these three levels, five testing techniques can be identified.

Desk Checking. This involves testing part of a system by tracing through it with a mental representation of a transaction. The testing can be undertaken at minimal cost, but the results may not be fully reliable because of the dependence upon human processing.

Random Data Testing. One-off tests can be performed on certain types of system which do not involve continuous processing by submitting representative data to the system. This data should invoke the possible transactions that the system will eventually perform and the effects observed.

Live Data Checking. An extension of random data testing is to test a system with *live* data, in which large samples of real transactions are input to the system. The advantage of this method is that it is relatively inexpensive and has a higher degree of reliability than the methods outlined above.

Production Testing. This involves the submission of all transactions to the system. The behaviour of the programs is tested, detected errors are corrected and transactions are resubmitted. This method of testing yields results of higher accuracy but the inevitable repetition of this process makes it relatively expensive.

Controlled Testing. Input is designed so as to allow for the processing of transactions covering all (or at least the majority) of the permutations. Thus, regular as well as irregular situations may be tested and therefore this method may result in fairly reliable results. However, the preparation of input may be expensive.

10.2 System Introduction

Once a system has been specified, designed, coded and tested, the final task remains its introduction. Of course, whilst such a task is the last in the development sequence, its planning and preparation will often begin long before coding commences.

In its broadest sense, *system introduction* is a collective term for a number of activities which must take place leading to a system becoming operational. These activities are the training and education of system users and the conversion from the old to new system.

10.2.1 Training and Educating Users

In training and educating users about a new system, two audiences must be addressed: operating and computer personnel and the actual users of the system, often known as *end-users*.

Computer personnel must be trained before the system *goes live* and is used by end-users but also training must be provided on a continuous basis as new personnel join the organisation. Training may be provided by the following means.

The Organisation. Training provided by the organisation is known as in-house training and is usually associated with large organisations who have established training departments.

Computer Manufacturers. The majority of computer manufacturers have training centres where customers can be trained on any aspect of computer usage, hardware, or software such as operating systems, database management systems, languages etc.

Training Firms. Training of generalised nature, such as in the programming language COBOL, may be obtained from a large number of training firms. This is an economically viable method of training for many small to medium-sized organisations, especially when a small number of staff are to be retrained.

Training for end-users is concerned with how users can obtain information from a system, as opposed to how the system operates. Typically, end-user training can be achieved through the use of short, in-house courses; dummy system operation by simulating the real environment and on the job training, in which end-users carry out particular tasks with the help of staff from the computing division.

10.2.2 System Conversion

The term conversion is used to describe the process of changing from one system to another, such as from a manual system to an automated system or from an existing automated system to a new one.

Typically, conversion planning needs to consider the following points:

- the collection and input of data, not currently used, for use by the new system

- the conversion of existing data into a form suitable for use by the system; this may require either conversion from a manual form to a computer readable form or the reformatting of existing computer files

- the manual procedures associated with the information system.

There are a number of different approaches towards accomplishing the conversion to a new system. These approaches are as follows.

Burning Bridges or Direct Approach. With the direct approach, conversion takes place all at once over a short period of time and requires the system to work correctly from the start, since there is no fall-back position. A major disadvantage of this approach is the high risk of failure. and therefore this approach is not widely used. There are, however, situations where the approach might be relevant, particularly when the new system is small or the old system has completely degenerated and has no value.

Pilot Approach. The pilot approach, sometimes referred to as the modular approach, undertakes conversion on a piecemeal basis. Only small parts of the total system are converted at a time.

The advantage of this method is that the effects of the new system relate to only a small part of the organisation. However, such an approach may lead to very lengthy implementation.

Phase-In/ Phase-Out Approach. The phased approach is one in which an old system is gradually replaced by a new one. In this sense it is similar to the pilot approach but the difference is that it is the computer-based information system which is segmented rather than the organisation itself.

The advantage of this method is the modular way of introducing a computer system. However, this modularity requires the specification of careful interfaces between the various models so that the system behaves in a coordinated manner.

Parallel Running. Parallel running is arguably the most popular method of system conversion. It involves the operating of the old and new system simultaneously for a period of time. The output of a new system is compared to that from the old. Any discrepancies initiate a chain of activities, usually re-programming a particular part of the new system, re-testing etc. The main advantage of this approach is that it provides a high degree of protection. The disadvantage is the cost of duplicating every computer related activity.

10.2.3 Behavioural Effects of Introducing a Computer System

The introduction of computer-based information systems into an organisation can drastically change the way many operations are carried out. An organisation may be affected, technically, economically and behaviourally. A change may be ideal in terms of a technical and economic sense but unless it accommodates human expectations and alleviates human fears, the change will seldom be successful.

Change may be good or bad. But whereas the appropriateness of a change may be easy to identify with regard to technical and economical considerations, this may not be so from the behavioural point of view. What may be regarded as good by some people, might be totally unacceptable to others.

An analyst must be aware of a number of strategic variables relating to the introduction of a change whether for a computer system or operating procedures. For example, it must be explained to people *why* a change is needed or wanted. The emotional barrier that people have against changing something that they are familiar with must also be regarded in trying to implement some change. Finally, the analyst must be aware that people perceive change differently.

In organisations the introduction of a computer-based information system affects both its formal and informal structures. Any information may change departmental boundaries and job descriptions (the formal structure) as well as social and group relations (the informal structure). If these factors are not considered and dealt with as the system is being developed, resistance in the form of aggression, projection or avoidance may result. The following factors are pertinent.

- Aggression is a form of attack on a system with the intent of making it either physically inoperative or ineffective.

- Projection is a way of blaming the system for any malfunction encountered while using or interacting with it.

- Avoidance is the withdrawal of people from interacting with the information system. This may be the result of frustration, such as that generated from not knowing how to input data to a computer. Avoiding a VDU therefore means avoiding frustration.

Avoiding or minimising user resistance is one of the most difficult problems in introducing a new system. Whilst a system developer knows what techniques and tools to use in order to develop an efficient system, he is not so well equipped with solving behavioural difficulties. However, the following points should be considered when a new system is introduced:

- consider how changes in the past were implemented
- discuss the system with all personnel or their representatives, who are directly affected
- set realistic goals.

10.3 Project Management

10.3.1 Introduction

Experience has shown that a successful system requires not only a coherent and well-defined approach to its technical development, but also requires careful co-ordination and management of that technique to ensure that manpower and development costs are contained and kept within budget.

A number of well-established techniques exist to assist in project management and these are outlined in the following sections. Their purpose can be summarised as follows:

- formulating estimations of time requirements
- monitoring progress
- comparing actual against planned performance.

10.3.2 Decision Criteria

Decision criteria encompass a number of differing techniques, for example, decision theory, maximin criteria, regret criteria. Each allows the user to place emphasis upon his own expectations of the market situation and thus decisions are made having taken note of the real possibilities of fluctuating markets. By using decision criteria, together with management experience, optimal courses of action may be calculated from the estimated figures provided. Hence conclusions as to the real viability of a project may be reached. It should be noted that other project management techniques may need to be applied to a project before any decision criteria are used, due to the requirement of an accurate estimation of total project cost, for use as input to the decision theories in use.

10.3.3 Milestone Charts

The use of milestone charts or project status reports involves the division of a project into separate, self-contained tasks. A table, or chart, is drawn up listing these tasks, together with relevant information. Each tuple of the table will contain such items as the task title, employee or team assigned, a review date and the completion date. Figure 10.1 illustrates part of a typical milestone chart.

PROJECT STATUS REPORT

System name:_____Hotel Project____________ Date:_____24.5.88______

Task number	Task description	Responsible		Man-days	Start	Completion
A1	Desk clerk task analysis	S.White	Plan	13	3.5.88	22.5.88
			Actual	16	2.5.88	24.5.88
			Diff.	1	1	2
A2	Define system boundary	A.Jones	Plan	4	4.5.88	9.5.88
			Actual	-	23.5.88	contin-uing
			Diff.	-	19	-
...		...		...		...

Figure 10.1: A Sample Milestone Chart

The milestone charts provide a convenient, easily prepared and understood reference table. A quick glance at the table will provide management with basic information about each modular task. However, while use of the chart will indicate whether a job has overrun its expected completion date, there is no way that this can be used to assess the overall effect that this will have on the project as a whole. A further limitation of milestone charts can be seen as their lack of provision for any other variables than time; resource scheduling or costs are not catered for.

10.3.4 Gantt Charts

Gantt charts, named after their originator, are very similar to milestone charts. Tasks are defined in the same manner and tabulated on the chart. Figure 10.2 illustrates such a chart.

The advantages, and disadvantages, of Gantt charts are similar to those specified for milestone charts.

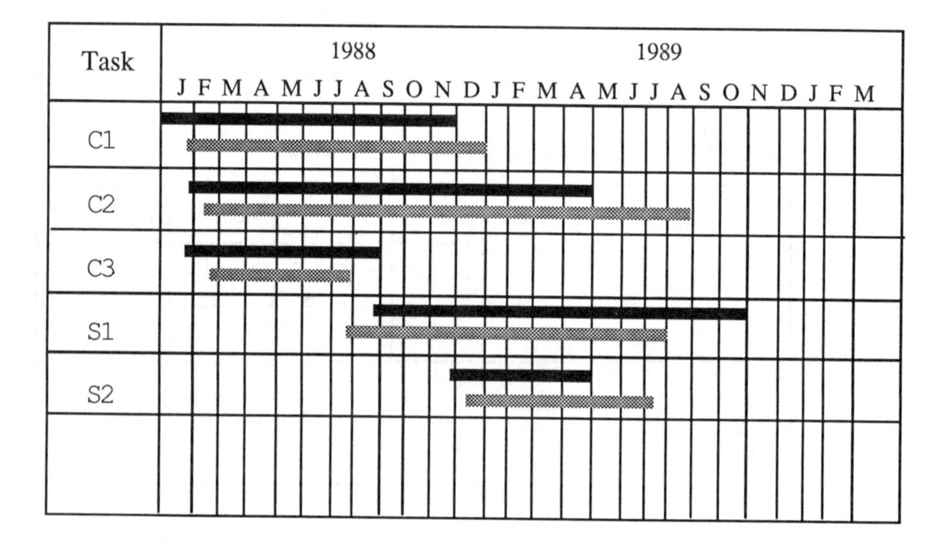

Figure 10.2: A Sample Gantt Chart

10.3.5 Networks

Networking is a generic name for a number of techniques which utilise what is termed *time network analysis* as a basis for the control of manpower, resources and capital used in a project.

The foundation of the approach came from the Special Projects Office of the US Navy in 1958. It developed a technique for evaluating the performance of large development projects, which became known as PERT- the Project Evaluation and Review Technique. Other variations of the same approach are known as the critical path method or critical path analysis (CPA) (*Beech & Burn, 1985*).

At the time of their conception, both PERT and CPA were virtually the same system. Both these project management techniques were concerned solely with the time-span of a project's tasks, their interrelation and, hence, the lifecycle of the project as a whole. However, over time, PERT has evolved to consider other project variables (*Saitow, 1969*).

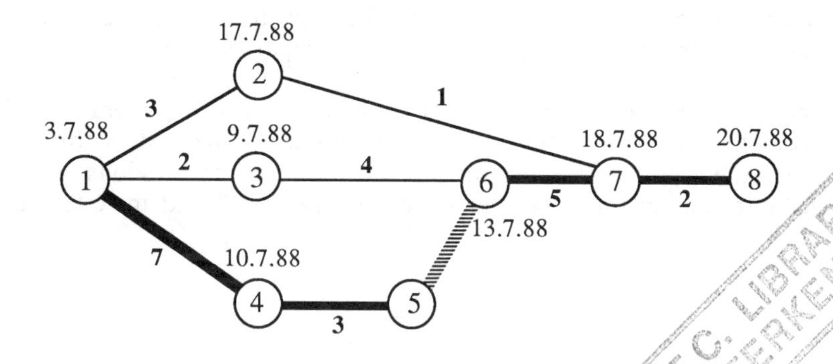

Figure 10.3: A Sample CPA Diagram

The heart of any PERT chart or CPA diagram is a network, such as that shown in figure 10.3. The diagram consists of a number of circles, representing events within the development lifecycle, such as the start or completion of a task and lines, which represent the tasks themselves. Each task is additionally labelled by its time duration. Thus the task between events 4 and 5 in figure 10.3 is planned to take 3 time units.

By careful construction of a CPA diagram, the interdependence of one task upon another is illustrated; that is for a task to be shown as being the predecessor of another. *Dummy* tasks, represented by dotted lines, are used to distinguish between the tasks associated with an event which has more than one dependent; these *dummy activities* are normally of zero duration.

Once a basic CPA diagram has been constructed, start and termination dates can be added.The dates shown in figure 10.3 represent the last date a task may start. Once these dates have been calculated, a *critical path* can be identified. This path, shown as a heavier line in figure 10.3, indicates those tasks which must be performed according to schedule if the overall end-date of the project is not to slip. Thus, the task between events 2 and 7 must start no later than 17.7.88 if it is not to delay other parts of the project.

10.4 Summary

In this chapter, a number of issues have been considered relating to ensuring that a system is delivered on-time, is introduced smoothly and operates error-free.

In summary, the important issues of project planning and control revolve around ensuring correct preparation before the introduction of a system. This preparation includes the proper testing of all aspects of a system, from individual program modules, through to the fully integrated system and in ensuring that users are adequately trained and prepared for a new system.

The other aspect of planning and control relates to overall project management, which although has been addressed at the end of this book, underpins the whole of the technical effort described in previous chapters. Further information on project management may be found in (*Cleland & King, 1983*).

References

Abbott, J. (1983) *Presentation of Computer I/O for People*, NCC Publications.

Adelson, B. & Soloway, E. (1985) *The Role of Domain Experience in Software Design*, in IEEE Trans. on Software Engineering, Vol.SE-11, No.11, November 1985.

Albano, A. et al (1985) *Galileo: A Strongly Typed, Interactive Conceptual Language*, ACM Trans.on Database Systems, Vol.10, No.2, June 1985.

AlFedaghi, S. & Scheuermann, P. (1981) *Mapping Considerations in the Design of Schemas for the Relational Model*, IEEE Transactions on Software Engineering, Vol.7, No.1, pp. 99-111 , 1981.

Allen J. (1983) *Maintaining Knowledge About Temporal Intervals*, Comm. ACM, Vol. 26, No. 11, Nov. 1983.

Alvey (1983), *The Alvey Report*, HMSO, UK.

ANSI (1975) *Interim Report of ANSI/X3/SPARC Group on Database Management Systems*, ANSI, February 1975.

Anthony, R.A. (1965) *Planning and Control Systems: A Framework for Analysis*, Division of Research, Graduate School of Business Administration, Harvard University.

van Assche, F., Layzell, P.J., Loucopoulos, P., Speltincx, G. (1988) *Information Systems Development : A Rule-Based Approach*, Journal of Knowledge Based Systems, September, 1988, pp. 227-234.

Balzer, R.M. & Goldman, N. (1979) *Principles of Good Software Specification and Their Implications for Specification Languages*, Proc. Spec. Reliable Software Conf., April 1979, pp. 58-67.

Balzer, R.M., Goldman, N., Wile, D. (1982) *Operational Specifications as the Basis for Rapid Prototyping*, in Proc 2nd Software Engineering Symposium: Workshop on Rapid Prototyping, ACM, New York, April 1982.

Batini,C. & Lenzerini,M. (1984) *A Methodology for Data Schema Integration in the Entity-Relationship Model,* IEEE Trans. on Software Engineering, Vol.SE-10, No.6, pp.650-664.

Beech, C. & Burn, J. (1985) *Application in Business Data Processing*, Pitman Press.

Bersoff, E.H., Henderson, V.D., Siegel, S.G. (1979) *Software Configuration Management- A Tutorial'*, IEEE Computer, January 1979.

Black, W.J., Sutcliffe A.G., Loucopoulos, P., Layzell, P.J. (1987) *Translation Between Pragmatic Software Development Methods*, European Software Engineering Conference, Strasburg, September 1987.

Blank, J. & Krijger, J. (1982) *Evaluation of Methods and Techniques for the Analysis, Design and Implementation of Information Systems*, Academic Press.

Boehm, B.W., Gray, T.E., Seewaldt, T. (1984) *Prototyping vs Specifying: A Multiproject Experiment*, IEEE Trans. on Software Engineering, Vol.SE-10, No.3, May 1984.

Bolour, A., Anderson, T.L., Dekeyser, L.J., Wong, H.K.T. (1982) *The Role of Time in Information Processing: A Survey*, ACM SIGART Newsletter, April 1982.

Borgida, A.T., Mylopoulos, J. & Wong ,H.(1982) *Methodological and Computer Aids for Interactive Information System Development*, in Automated Tools for Information Systems Design, Schneider and Waserman (eds), North-Holland Publishing Company.

Borgida, A. (1986) *Survey of Conceptual Modelling of Information Systems*, in On Knowledge Base Management Systems, Brodie, M.L. & Mylopoulos, J. (eds), Springer-Verlag, 1986.

Bourginon, J-P (1986) *PCTE - A Portable Common Tool Environment*, ESPRIT Technical Week, Brussels, September 1986.

Brachman, R. & Levesque, H. (1984) *The Knowledge Level of KBMS*, in On Conceptual Modelling: Perspectives from Artificial Intelligence, Databases, and Programming Languages, Brodie, M.L., Mylopoulos, J. (eds), Springer-Verlag, 1984.

Brodie, M.L. & Silva E. (1982) *Active and Passive Component Modelling (ACM/PCM)*, in Information Systems Design Methodologies: A Comparative Review, Olle,T.W et al (eds), IFIP TC8, North-Holland.

Brodie, M.L., Mylopoulos, J., Schmidt, J.W. (eds) (1984) *On Conceptual Modelling: Perspectives from Artificial Intelligence, Databases, and Programming Languages*, Springer- Verlag, 1984.

Bubenko, J. (1980) *Information Modelling in the Context of System Development*, in Proc. IFIP 1980, North-Holland.

Cardenas, A.F. (1973) *Evaluation and Selection of File Organisations - A Model and a System*, Comm. ACM, Vol.16.

Casanova,M.A., & Vidal,V.M.P. (1983) *Towards a Sound View Integration Methodology,* Proceedings of the 2nd ACM SIGACT/SIGMOD Conference on Principles of Database Systems. Atlanta, Georgia, May 21-23, pp.36-47,1983.

Champine, G.A. (1980) *Distributed Computer Systems*, Impact on Management, Design and Analysis, North-Holland.

Chapin, N. (1979) *Software Lifecycle*, INFOTEC Conf. on Structured Software Development.

Chen, P.P. (1976) *The Entity-Relationship Model: Towards a Unified View of Data*, in ACM Transactions on Database Systems, Vol.1, No.1, March 1976, pp.9-36.

Chen, P.P. (1977) *The Entity-Relationship Model- A Basis for the Enterprise View of Data*, National Computer Conference.

Clare, C.P. & Loucopoulos, P. (1983) *Data Processing: Current Theories and Practices*, Abacus Press.

Cleland, D.I. & King, W.R. (1983) *Systems Analysis and Project Management*, McGraw-Hill.

Clemons, E K (1985) *Data Models and the ANSI/SPARC architecture*, in Principles of Database Design, Yao S.B (ed), Prentice Hall.

Clocksin, W.F. & Mellish, C.S. (1984) *Programming in Prolog*, Springer-Verlag.

Cobb, R.H. (1985) *In Praise of 4GLs*, Datamation, July 1985.

CODASYL (1971) CODASYL Data Base Task Group Report, Conf. on Data System Languages, ACM, New York.

CODASYL (1973) CODASYL *Data Description Language Journal of Development*, National Bureau of Standards Handbook 113. U.S. Government Printing Office, (SD Catalogue No C13.6/2:113), Washington, DC.

CODASYL (1976) CODASYL *COBOL Journal of Development*, Material Data Management Branch, Dept of Supply and Services, Ottawa.

CODASYL (1977) CODASYL *FORTRAN, Data Base Facility Journal of Development* Material Data Management Branch, Dept of Supply & Services, Ottawa.

CODASYL (1978) CODASYL *Data Description Language Journal of Development*, Material Data Management Branch, Dept of Supply & Services, Ottawa.

Codd, E.F. (1970) *A Relational Model of Data for Large Shared Data Banks*, Comm. ACM., Vol.13, No.6.

Codd, E.F. (1972) *Further normalisation of the data base relational model*, Data Base Systems, Courant Computer Symposia Series, Vol.6, Prentice Hall.

Codd, E.F. (1979) *Extending The Database Relational Model to Capture More Meaning*, ACM Trans. on Database Systems, Vol.4, No.4, December 1979.

Couger, J.D., Colter, M.A. & Knapp, R.W. (1982) *Advanced System Development/Feasibility Techniques*, J.Wiley.

Curtis, B., Krasner, H., Iscoe, N. (1988) *A Field Study of the Software Design Process for Large Systems*, Comm. ACM, Vol.31, No.11.

Date, C.J. (1981) *An Introduction to Database Systems*, Addison-Wesley.

Davis, G.B. (1974) *Management Information Systems: A Framework for Planning and Development*, McGraw-Hill.

Davenport, R.A. (1981) *Design of distributed database systems*, BCS Computer Journal, Vol.24, No.1.

Dean, T. (1987) *Large Scale Temporal Databases for Planning in Complex Domains*, Proc 10th IJCAI, Milan, August 1986.

de Marco, T. (1978) *Structured Analysis and System Specification*, Yourdon Press.

De Troyer, O. & Meersman, R. (1986) *Transforming Conceptual Schema Semantics to Relational Data Applications*, in Information Modelling and Database Management Systems, Springer-Verlag.

Dubois E., Hagelstein, J., Lahou, E., Rifaut, A. & Williams, F. (1986) *A Data Model for Requirements Engineering*, Proc. 2nd Int. Conf. on Data Engineering, Los Angeles, pp.646-653.

El Masri, R., Larson, J. and Navathe, S.B. (1987) *Integration Algorithms for Federated Databases and Logical Database Design*, Technical Report, Honeywell Corporate Research Center, 1987.

ESPRIT (1985) *ESPRIT Programme of Research and Development 1985.*

Floyd, C. (1986) *A Comparative Evaluation of System Development Methods*, in Information Systems Design Methodologies: Improving the Practice, Olle,T.W et al (eds), IFIP TC8, North-Holland.

Fickas, S. (1987) *Automating the Analysis Process: An Example*, 4th Int. Workshop on Software Specification & Design, Monterey, USA.

Flynn, D.J., Layzell, P.J., Loucopoulos, P. (1986) *Assisting the Analyst - The Aims and Objectives of the Analyst Assist Project*, BCS Software Engineering 86, Southampton.

Galton, A. (ed) (1987) *Temporal Logics and their Applications*, Academic Press, 1987.

Gane, C. & Sarson, T. (1979) *Structured Systems Analysis: Tools and Techniques*, Prentice Hall.

Gray, J. (1981) *The Transaction Concept: Virtues and Limitations*, in Readings in Database Systems, Stonebraker, M. (ed), Morgan Kaufmann Publishers Inc, 1988, pp.140-149.

Greenspan, S.J. (1984) *Requirements Modeling: A Knowledge Representation Approach to Software Requirements Definition*, Technical Report No. CSRG-155, University of Toronto, 1984.

Greenspan, S.J. & Mylopoulos, J. (1982) *Capturing More World Knowledge in the Requirements Specification*, Proc. 6th Int. Conf. on Software Engineering, Tokyo, pp.225-234.

van Griethuysen, J.J. et al (eds). (1982) *Concepts and Terminology for the Conceptual Schema and the Information Base*, Report ISO, TC97/SCS/WG3.

Griffiths, S.N. (1978) *Design Methodologies - A Comparison*, Pergamon Infotech, 1978, pp.133-166.

Gross, J. et al (1980) *Distributed Database: Design and Administration*, Data Logic.

Gustafsson, M.R., Karlsson, T., Bubenko, J.A. (1982) *A Declarative Approach to Conceptual Information Modelling*, in Information Systems Design Methodologies: A Comparative Review, Olle,T.W et al (eds), IFIP TC8, North-Holland.

Hagelstein, T. (1988) *Declarative Approach to Information Systems Requirements*, Knowledge Based Systems, Vol.1, No.4, pp.211-220.

Hall, J. (1981) *System Development Methodology*, Learmonth & Burchett Management Systems, London.

Hanson, O. (1978) *Basic File Design*, IPC Business Press.

Hanson, O. (1977) *Minimising the Effects of Synonyms on the Processing Time of Directly Organised Files*, Proc. 5th Int. Congress on Data Processing in Europe, Vienna.

Hirscheim, R.A. (1981) *Information Management in Organisations*, Working Paper wp-4-81, London School of Economics, 1981.

Howe, E.R. (1983) *Data Analysis for Database Design*, Edward Arnold.

Hsiao, D.K. & Harary F (1970) *A Formal System for Information Retrieval from Files*, Comm. ACM, Vol.13.

Jackson, M.A. (1975) *Principles of Program Design*, Academic Press.

Jackson, M.A. (1983) *System Development*, Prentice Hall.

Jardine, D.A., Matzov, A. (1986) *Ontology and Properties of Time in Information Systems*, Proc. of the Conference on Knowledge and Data, Portugal, 1986.

Jarke, M., Jeusfeld, M., Rose, T. (1988) *Modelling Software Processes in a Knowledge Base: The Case of Information Systems*, Knowledge Based Systems, Vol.1, No.4, pp.197-210.

Johnson, P., Diaper, D., Long, J. (1984) *Tasks, skills and knowledge: Task analysis for knowledge-based descriptions*, Interact '84, 1st IFIP Conference on Human-Computer Interaction, Vol.1, pp.23-27.

Jones, C.B. (1981) *Software Development: A Rigorous Approach*, Prentice Hall, London, 1981.

Kahn, B.K. 1979) *A Structured Logical Data Base Design Methodology*, Ph.D dissertation, Computer Science Department, University of Michigan, Ann Arbor, Michigan, 1979.

Karakostas, V. & Loucopoulos, P. (1988) *Verification of Conceptual Schemata Based on a Hybrid Object Oriented and Logic Paradigm*, Journal of Information and Software Technology, Vol.30, No.10, pp.587-594.

Kent, W. (1983) *A Simple Guide to Five Normal Forms in Relational Database Theory*, in Comm. ACM, Vol.26, No.2, February 1983.

Kent, W., (1983), *The Realities of Data: Basic Properties of Data Reconsidered*, in Proc. IFIP WG 2.6 Working Conference on Data Semantics (DS-1), Hasselt, Belgium, 7-11 January, 1985, North-Holland, 1986, pp.175-188.

Kerola, P., (1988) *On the Necessary Levels of Abstraction in the Framework of Information System Concepts*, IFIP WG 8.1 meeting, January 17-19, 1988, Barcelona.

Klein, H.K. & Hirschheim, R.A. (1987) *A Comparative Framework of Data Modelling Paradigms and Approaches*, The Computer Journal, Vol.30, No.1, pp. 8-15.

Land, F. F. (1980) *Adapting to changing User Needs*, Infotech State of the Art Report, in Life Cycle Management, Series No. 8.

Land, F. F. (1981) *Concepts and Perspectives: A Review*, in Proc. IFIP TC8 Working Conference on Evolutionary Information Systems, Budapest, Hungary, 1-3 September.

LBMS (1987) *Auto-Mate Plus Manuals*, Volumes 1, 2 & 3, LBMS plc, London.

Lee, R. (1986) *Logic, Semantics and Data Modelling: An Ontology*, in Conference pre-prints, IFIP TC2 Working Conference, 'Knowledge and Data', Algarve, Portugal, November 3-7, 1986.

Lehman, M. (1978) *Software Engineering and the Characteristics of Large Programs*, Infotec.

Lientz, B.P. & Swanson, E.B. (1980) *Software Maintenance Management*, Addison Wesley.

Longworth, G. & Nicholls, D. (1986) *SSADM Manual*, Volumes 1 & 2, National Computing Centre, Manchester, UK.

Loucopoulos, P., Black, W.J, Sutcliffe, A.G., Layzell, P.J.(1987) *Towards a Unified View of System Development Methods*, International Journal of Information Management, Vol.7, No.4, Butterworths.

Lundeburg, M., Goldkuhl, G. & Nilssen A. (1981) *Information Systems Development: A Systematic Approach*, Prentice Hall.

Macdonald, I.G. (1986) *Information Engineering: An Improved, Automated Methodology for the Design of Data Sharing Systems*, in Information Systems Design Methodologies: Improving the Practice, Olle,T.W et al (eds), IFIP TC8, North-Holland.

Maddison, R.N. et al (1983) *Information System Methodologies*, Wiley Heyden.

Marchetti, C. (1981) *Society as a Learning System: Discovery, Invention and Innovation cycles Revisited*, in Market & Product Innovation Facing Social & Technological Change, Turin, April, 1981.

Martin, J. (1983) *An Information Processing Manifesto*, Savant, Carnforth, UK.

Martin, J. & McClure, C (1984) *Structured Techniques for Computing*, Savant Institute.

Mayhew, P.J. & Dearnley, P.A. (1987) *An Alternative Prototyping Classification*, Computer Journal, Vol.30, No.6, 1987.

McDermott, D. (1982) *A Temporal Logic For Reasoning About Processes and Plans*, Cognitive Science 6, December 1982.

Moser, C.A & Kalton, G. (1971) *Survey Methods in Social Investigation*, Heinemann Educational Books.

Mumford, E. et al (1978) *A Participative Approach to the Design of Computer Systems*, Impact of Science on Society, Vol.28, No.3.

Myers, G.J. *Reliable Software Through Composite Design*, Petrocelli/Charter.

Mylopoulos, J. (1986) *The Role of Knowledge Representation in the Development of Specifications*, In Information Processing 86, Kugler, H-J. (ed), Elsevier Science Publishers.

Mylopoulos, J., Bernstein, P.A., Wong ,H.K.T. (1980) *A Language Facility for Designing Database-Intensive Applications*, ACM Transactions on Database Systems, Vol. 5, No. 2, June 1980, pp.185-207.

Navathe, S. B, & Gadgil, S. G. (1982) *A Methodology for View Integration in Logical Database Design.* Proceedings of the 8th International Conference on Very Large Data Bases, Mexico City, pp.142-164.

Newell, A. & Simon, H.A. (1972) *Human Problem Solving*, Prentice Hall.

Nijssen, G.M (1977) *Architecture and Models in Data Base Management Systems*, Proc IFIP Working Conference on Modelling in Data Base Management Systems, Nice, France.

Nijssen, G.M. & Duke, D.J. (1987) *Real and Information Systems Concepts*, Position paper for IFIP WG 8.1 (FRISCO), Copenhagen, 15-16 October 1987.

Nijssen, G.M., Duke, D.J., Twine, S.M., (1988), *The Entity-Relationship Data Model Considered Harmful*, Proc 6th Symposium on Empirical Foundations of Information & Software Sciences, Atlanta, Georgia, USA, October 1988.

Ogden, C.K. & Richards, I.A. (1923) *The Meaning of Meaning*, Harcourt, New York.

Olivé, A. (1983) *Analysis of Conceptual and Logical Models in Information Systems Design Methodologies*, In Information Systems Design Methodologies: A Feature Analysis, Olle, T., Sol, H., and Tully, C. (eds), North Holland Publ. Co. 1983.

Olle, T.W. (1988) *Orthogonal Concepts and Useful Pairings for Information Systems*, IFIP WG 8.1 meeting, January 17-19, 1988, Barcelona.

Olle, T.W., Sol, H.G. & Verrijn Stuart, A.A. (1982) *Information Systems Design Methodologies: A Comparative Review*, IFIP TC8, North-Holland.

Olle, T.W., Sol, H.G. & Tully C.J. (1983) *Information Systems Design Methodologies: A Feature Analysis*, IFIP TC8, North-Holland.

Oxborrow, E. (1986) *Databases and Database Systems*, Chartwell-Bratt.

Reynolds, C.H. (1978) *Issues in Centralisation*, Datamation, March 1978.

Rock-Evans, R. (1981) *Data Analysis*, IPC Business Press.

Roman, G-C. (1985) *A Taxonomy of Current Issues in Requirements Engineering*, IEEE Computer, April 1985, pp.14-22.

Ross, D.T. & Schoman, K.E. (1977) *Structured Analysis (SA): A Language for Communicating Ideas*, IEEE Trans. on Software Engineering, Vol.SE-3, No.1, January 1977, pp.16-34.

Royce, W. (1970) *Managing the Development of Large Software Systems*, Proc. IEEE Wescon, August 1970, pp.1-9.

Rzepka, W. & Ohno, Y. (1985) *Requirements Engineering: Software Tools for Modelling User Needs* , IEEE Computer, April 1985, pp.9-12.

Saitow, A. (1969) *CSPC: Reporting Project Progress to the Top*, Harvard Business Review.

Sakai, H. (1981) *A method for defining information structures and transactions in conceptual schema design*, ACM SIGMOD.

Schank, R.C. (1975) *The Structure of Episodes in Memeory*, in Representation & understanding, studies in Cognitive Science, (ed Bobrow & Collins), Academic Press, New York.

Schank, R.C. (1976) *Conceptual Information Processing*, North-Holland, Amsterdam.

Severance, D.G. (1975) *A Parametric Model of Alternative File Structures*, Information Systems 1.

Simon, H.A (1960) *The New Science of Management Decisions*, Harper & Row.

Sowa, J.F. (1984) *Conceptual Structures: Information Processing in Mind and Machine*, Addison-Wesley Publishing Company, 1984.

Stevens, W.P, Myers, G.J. & Constantine, L.L (1974) *Structured Design,* IBM Systems Journal, No.2, pp.115-139.

Talbot, D.E. (1985) *Current Developments in Software Engineering Relevant to Data Processing*, Alvey/BCS SGES Workshop, Sunningdale, UK, Jan 1985.

Tardieu, H. (1981) *MERISE,* Journees Internationales de L' informatique et de L' automatique, Paris.

Tardieu, H.(1985) *MERISE, demarce et pratiques,* Paris, Les editions d'organisation, 1985.

Taylor, J. (1975) *The Human Side of Work: The Socio-Technical Approach to Work Design*, Personnel Review, Vol.4, No.3.

Teorey,T.J. & Fry,J.P. (1982) *Design of data base structures*, Prentice Hall.

Triance, J.M. & Edwards, B.J. (1979) *A computer aided program design project*, in Proc. Euro IFIP-79 Conference pp.621-626

TSE (1977) *IEEE Transactions on Software Engineering special issue on Requirments Specification*, Vol. SE-3, No. 1.

Tsichritzis, D.C. & Lochovsky, F.H. (1983) *Data Models*, Prentice Hall.

Tsichritzis, D.C. & Lochovsky, F. H. (1977) *Database Management Systems*, Academic Press.

Ullman, J.D. (1980) *Principles of Database Systems*, Computer Science Press.

Verheijen, G. & van Bekkum, J. (1982) *NIAM: An Information Analysis Method* , in Information Systems Design Methodologies: A Comparative Review, Olle,T.W et al (eds), IFIP TC8, North-Holland.

Verrijn-Stuart, A.A. (1987) *Themes and Trends in Information Systems*, Computer Journal, Vol.30, pp.97-109.

Verrijn-Stuart, A.A. (1988) *Paradigms, Theories and Concepts*, FRISCO position paper, IFIP WG.8.1 meeting, Barcelona, Spain, 16-19 January 1988.

Vetter, M. & Maddison, R. (1981) *Database Design Methodology*, Prentice Hall.

Vitalari, N.P. & Dickson, G.W. (1983) *Problem Solving for Effective Systems Analysis: An Experimental Exploration*, Comm. ACM, Vol.26, No.11, November 1983.

Wiederhold, G., Fries, J.F., Weyl, S. (1975) *Structured Organization of Clinical Databases*, in Proc of the NCC, AFIPS Press, Montvale, New Jersey.

Wiederhold, G. & El Masri, R. (1979) *A Structural Model for Database Systems*. Rep. STAN-CS-79-722, Computer Science Department, Stanford University, Stanford, California, 1979.

Winston, P.H. & Horn, B.K.P. (1984) *LISP*, Addison Wesley.

Yao, S. B., Waddle, V. E. and Housel, B. C. (1982) *View Modelling and Integration Using the Functional Model*. IEEE Trans.on Software Engineering, Vol.SE-8, No 6, pp.544-553.

Yeh, R.T. (1982) *Requirements Analysis - A Management Perspective*, Proc. COMPSAC '82, pp.410-416.

Yourdon, E. (1977) *Structured Walkthroughs*, Prentice Hall.

Yourdon, E. & Constantine, L.L. (1975) *Structured Design*, Yourdon Press.

Appendix A

Method Reference and Bibliography

Method	Reference
ACM/PCM	Brodie, M.L., *Specification and Verification of Database Semantic Integrity*, CSRG-91, University of Toronto, 1978.
	Brodie, M.L. & Silva, E. *Active and Passive Component Modelling*, in Information Systems Design Methodologies: A Comparative Review, Olle,T.W et al (eds), pp.41-91, IFIP TC8, North-Holland, 1982.
BC	Martin J., *Computer Data Base Organisation*, Prentice Hall, 1975.
BSP	IBM, *Business Systems Planning*, in Advanced Systems Development/ Feasibility Techniques, Cougar J.D., Colter, M.A. & Knapp, R.W., John Wiley, 1982.
CIAM	Bubenko, J.A., *Inferential Abstract Modelling,* IBM T.J. Watson Research Center, Yorktown Heights, NY. 10598, RC 6343, January 1977.
	Bubenko, J.A., *Information Modelling in the Context of System Development*, IFIP Congress, Tokyo and Melbourne, 1980.

D2S2 Palmer, I.R., *Practicalities in Applying a Formal Methodology to Data Analysis,* in Data Analysis for Information System Design, Maddison, R.N., British Computer Society, London, 1978.

Palmer, I.R., *System Development in a Shared Data Environment, the DSS Methodology,* in Information Systems Design Methodologies: A Comparative Review, Olle,T.W et al (eds), IFIP TC8, North-Holland, 1982.

DADES Olive, A. & Saltor, F. *Formal Verification of Information Derivability in Databases Using Precedence Analysis.* Technical report, Facultat d'Informàtica, Universitat Politècnica de Barcelona, 1981.

EDM Rzevski, G., *On the Design of a Design Methodology,* in Design, Science, Method, Jacques, R. & Powell, J.A. (eds), IPC Business Press, 1981.

HIPO *IBM HIPO, A Design Aid and Documentation Technique,* GC20-1851, IBM.

HOS Hamilton, M. & Zeldin, S., *Integrated Software Development System/Higher Order Software Conceptual Description,* TR-3, Higer Order Software Inc., Cambridge, Massachusetts, 1976.

IE Martin, J. & Finkelstein, C., *Information Engineering,* Volumes 1 & 2, Prentice Hall, 1981.

IML Richter, G., *Utilization of Data Access and Manipulation in Conceptual Schema Definitions,* Information Systems, Vol.6, No.1, pp.53-71, Pergamon Press, 1981.

Richter, G., *IML Inscribed Nets for Modeling Text Processing and Database Management Systems,* in Proc. 7th Int. Conf. on Very Large Databases, Cannes, pp.363-375, Morgan Kaufmann, 1981.

ISAC Lundeberg M. et al, *A Systematic Approach to Information Systems Development,* Information Systems, Vol.4, pp.1-12, 93-118, Pergamon Press Ltd. 1979.

ISSM

Sølvberg, A., *A Contribution to the Definition of Concepts for Expressing Users' Information Systems Requirements*, in P.P. Chen(ed): Entity-Relationship Approach to Systems Analysis & Design, North-Holland, 1980.

Stavelin, S., *Derivabilty Analysis and Data Modelling*, Masters thesis, Dept. of Computer Science, Norwegian Institute of Technology, Trondheim, Norway, 1980.

JSD

Jackson, M., *System Development*, Prentice Hall, 1983.

MERISE

Introduction a MERISE, journées internationales de L'informatique et de L'automatique, Paris, 1981.

Tardieu H., *MERISE,* in Information & Management, Amsterdam, 1983.

NIAM

Verheijen, G. & van Bekkum, J., *NIAM: An Information Analysis Method*, in Information Systems Design Methodologies: A Comparative Review, Olle,T.W et al (eds), pp.588-589, IFIP TC8, North-Holland, 1982.

PSL/PSA

Teichroew, D. & Hershey, E.A., *PSL/PSA: A Computer-Aided Technique for Structured Documentation and Analysis of Information Processing Systems*, IEEE Trans.on Software Engineering, Vol.SE-3, No.1, January 1977.

REMORA

Rolland, C. & Foucaut, O., *Concepts of the Design of an Information System Conceptual Schema and its Utilisation in the Remora Project,* Proc. 4th Int.Conf. on Very Large Databases, Berlin, Morgan Kaufmann, 1978.

SADT

Ross, D.T. & Schoman, K.E., *Structured Analysis (SA): A Language for Communicating Ideas*, IEEE Trans.on Software Engineering, Vol.SE-3, No.1, January 1977.

Conner, M.F., *Structured Analysis and Design Technique*, Softech Report #9595-7, Softech Inc., Waltham Mass. U.S.A..

SASD

De Marco, T., *Structured Analysis and System Specification,* Yourdon Press, 1978.

Yourdon, E.N. & Constatine, L.L., *Structured Design*, Yourdon Press & Prentice Hall, 1978.

SDLA

Knuth, E., Radó, P., Tóth, A., *Preliminary Description of SDLA*, Studies 105/1980, Computer & Automation Institute Hungarian Academy of Science, 1979.

SREM

Alford, M., *A Requirements Engineering Methodology for Real-Time Processing Requirements*, IEEE Trans. Software Engineering, Vol.SE-3, No.1, January 1977, pp.366-374.

SSA

Gane C. & Sarson T., *Structured System Analysis*, Prentice Hall, 1983.

SSADM

Longworth G. & Nicholls D., *SSADM Manual*, Volumes 1 & 2, National Computing Centre, 1986.

Downs, E. et al, *Structured Systems Analysis and Design Method: Application and Context*, Prentice Hall, 1988.

SYSDOC

Aschim, F. & Mostue, B.M., *Case Solved Using SYSDOC and Systemator*, in Information Systems Design Methodologies: A Comparative Review, Olle,T.W et al (eds), pp.15-40, IFIP TC8, North-Holland, 1982.

USE

Wasserman, A.I., *USE: A Methodology for the Design and Development of Interactive Information Systems,* in Formal Models & Practical Tools for Information System Design, Schneider, H.J. (ed.), North-Holland, pp.31-50.

Warnier Orr

Kenneth T. Orr, *Structured Systems Development*, Yourdon Press, 1977.

Higgens, D., *Program Design and Construction*, Prentice Hall, Englewood Cliffs, 1979.

Appendix B

Hotel Case Study

This appendix presents a simple case study. Its purpose is to serve as a common basis for demonstrating, by example, the concepts and techniques detailed in the main body of this book. The case study refers to an organisation that owns a number of hotels at various resorts. Rather than describing every detail of this system in this section, we concentrate on its top-level description. This top-level description is elaborated in the book, where appropriate.

The main components of the hotel organisation are as follows.

Customer Request

A person who wishes to go on a holiday makes a request to the company. The request may be very exact consisting of a specific resort, a specific date, a particular type of room in a specified hotel class. Alternatively, a request may be less exact, for example a customer may be interested only in the resort and the date. A request results in an offer being made to the customer which may or may not match the customer's requirements or it cannot be satisfied and is postponed.

Room Reservation

A reservation is made in an hotel for a customer in a given date, for a given period of time as a result of the customer accepting the offer made by the company. A reservation is provisional until the customer actually checks in the hotel, but a reservation can be cancelled at any time between acceptance and checking in. A person may only check in for a previously reserved room.

Invoicing

Invoicing involves two aspects. Firstly, a customer is invoiced when a room is booked and this constitutes the standard charge before the customer checks in. Secondly, a customer is invoiced when he checks out of the hotel. This relates to the charges which may be incurred during the customer's stay at the hotel. A customer must pay his bill on departure.

Index

Linear Programming: A Computational Approach: 2nd Ed, K K Lau, 150pp, ISBN 0-86238-182-7
Programming for Beginners: the structured way, D Bell & P Scott, 178pp, ISBN 0-86238-130-4
Software Engineering for Students, M Coleman & S Pratt, 195pp, ISBN 0-86238-115-0
Software Taming with Dimensional Design, M Coleman & S Pratt, 164pp, ISBN 0-86238-142-8
Systems Programming with JSP, B Sanden, 186pp, ISBN 0-86238-054-5

MATHEMATICS AND COMPUTING

Fourier Transforms in Action, F Pettit, 133pp, ISBN 0-86238-088-X
Generalised Coordinates, L G Chambers, 90pp, ISBN 0-86238-079-0
Statistics and Operations Research, I P Schagen, 300pp, ISBN 0-86238-077-4
Teaching of Modern Engineering Mathematics, L Rade (ed), 225pp, ISBN 0-86238-173-8
Teaching of Statistics in the Computer Age, L Rade (ed), 248pp, ISBN 0-86238-090-1
The Essentials of Numerical Computation, M Bartholomew-Biggs, 241pp, ISBN 0-86238-029-4

DATABASES AND MODELLING

Database Analysis and Design, H Robinson, 378pp, ISBN 0-86238-018-9
Databases and Database Systems, E Oxborrow, 256pp, ISBN 0-86238-091-X
Data Bases and Data Models, B Sundgren, 134pp, ISBN 0-86238-031-6
Text Retrieval and Document Databases, J Ashford & P Willett, 125pp, ISBN 0-86238-204-1
Towards Transparent Databases, G Sandstrom, 192pp, ISBN 0-86238-095-2
Information Modelling, J Bubenko (ed), 687pp, ISBN 0-86238-006-5

UNIX

An Intro to the Unix Operating System, C Duffy, 152pp, ISBN 0-86238-143-6
Operating Systems through Unix, G Emery, 96pp, ISBN 0-86238-086-3

SYSTEMS ANALYSIS AND DEVELOPMENT

Systems Analysis and Development: 3rd Ed, P Layzell & P Loucopoulos, 284pp, ISBN 0-86238-215-7

SYSTEMS DESIGN

Computer Systems: Where Hardware meets Software, C Machin, 200pp, ISBN 0-86238-075-8
Distributed Applications and Online Dialogues: a design method for application systems, A Rasmussen, 271pp, ISBN 0-86238-105-3

HARDWARE

Computers from First Principles, M Brown, 128pp, ISBN 0-86238-027-8
Fundamentals of Microprocessor Systems, P Witting, 525pp, ISBN 0-86238-030-8

NETWORKS

Communication Network Protocols: 2nd Ed, B Marsden, 345pp, ISBN 0-86238-106-1
Computer Networks: Fundamentals and Practice, M D Bacon *et al,* 109pp, ISBN 0-86238-028-6
Datacommunication: Data Networks, Protocols and Design, L Ewald & E Westman, 350pp, ISBN 0-86238-092-8
Telecommunications: Telephone Networks 1, Ericsson & Televerket, 147pp, ISBN 0-86238-093-6
Telecommunications: Telephone Networks 2, Ericsson & Televerket, 176pp, ISBN 0-86238-113-4

GRAPHICS

An Introductory Course in Computer Graphics, R Kingslake, 146pp, ISBN 0-86238-073-1
Techniques of Interactive Computer Graphics, A Boyd, 242pp, ISBN 0-86238-024-3
Two-dimensional Computer Graphics, S Laflin, 85pp, ISBN 0-86238-127-4

APPLICATIONS

Computers in Health and Fitness, J Abas, 106pp, ISBN 0-86238-155-X
Developing Expert Systems, G Doukidis, E Whitley, ISBN 0-86238-196-7
Expert Systems Introduced, D Daly, 180pp, ISBN 0-86238-185-1
Handbook of Finite Element Software, J Mackerle & B Fredriksson, approx 1000pp, ISBN 0-86238-135-5
***Inside* Data Processing: computers and their effective use in business,** A deWatteville, 150pp, ISBN 0-86238-181-9
Proceedings of the Third Scandinavian Conference on Image Analysis, P Johansen & P Becker (eds) 426pp, ISBN 0-86238-039-1
Programmable Control Systems, G Johannesson, 136pp, ISBN 0-86238-046-4
Risk and Reliability Appraisal on Microcomputers, G Singh, with G Kiangi, 142pp, ISBN 0-86238-159-2
Statistics with Lotus 1-2-3, M Lee & J Soper, 207pp, ISBN 0-86238-131-2

HCI

Human/Computer Interaction: from voltage to knowledge, J Kirakowski, 250pp, ISBN 0-86238-179-7
Information Ergonomics, T Ivegard, 228pp, ISBN 0-86238-032-4
Computer Display Designer's Handbook, E Wagner, approx 300pp, ISBN 0-86238-171-1

INFORMATION AND SOCIETY

Access to Government Records: International Perspectives and Trends, T Riley, 112pp, ISBN 0-86238-119-3
CAL/CBT - the great debate, D Marshall, 300pp, ISBN 0-86238-144-4
Economic and Trade-Related Aspects of Transborder Dataflow, R Wellington-Brown, 93pp, ISBN 0-86238-110-X
Information Technology and a New International Order, J Becker, 141pp, ISBN 0-86238-043-X
People or Computers: Three Ways of Looking at Information Systems, M Nurminen, 1218pp, ISBN 0-86238-184-3
Transnational Data Flows in the Information Age, C Hamelink, 115pp, ISBN 0-86238-042-1

SCIENCE HANDBOOKS

Alpha Maths Handbook, L Rade, 199pp, ISBN 0-86238-036-7
Beta Maths Handbook, L Rade, 425pp, ISBN 0-86238-140-1
Handbook of Electronics, J de Sousa Pires, approx 750pp, ISBN 0-86238-061-8
Nuclear Analytical Chemistry, D Brune *et al,* 557pp, ISBN 0-86238-047-2
Physics Handbook, C Nordling & J Osterman, 430pp, ISBN 0-86238-037-5
The V-Belt Handbook, H Palmgren, 287pp, ISBN 0-86238-111-8

Chartwell-Bratt specialise in excellent books at affordable prices.

For further details contact your local bookshop, or ring Chartwell-Bratt dircct on **01-467 1956** (Access/Visa welcome.)

Ring or write for our *free* catalogue.

Chartwell-Bratt (Publishing & Training) Ltd, Old Orchard, Bickley Road, Bromley, Kent, BR1 2NE, United Kingdom.
Tel 01-467 1956, Fax 01-467 1754, Telecom Gold 84:KJM001, Telex 9312100451(CB)

Software Engineering for Students

BY MICHAEL COLEMAN AND STEPHEN PRATT

Software Engineering for Students is an introductory text suitable for anyone studying programming or software engineering. It is divided into three sections:-

1. Introduction Covers the need for software engineering and for associated documentation.

2. Design Issues Concentrates on requirements specification, data analysis, design methodologies, verification and testing.

3. Implementation Issues Deals with programming languages, tools and techniques for software development, testing and debugging, ending with a look at the future for software engineering.

Providing a host of examples in BBC **BASIC** and **Pascal**, Software Engineering for Students clearly illustrates the principles with which every student, and software engineer, should be familiar.

Stephen Pratt is Principal Lecturer in the School of Information Science at Portsmouth Polytechnic. Michael Coleman, who has previously held that position, is currently employed by IBM.

195 pages, ISBN 0-86238-115-0

CURRENT PRICE £5.95 (confirm details before ordering)

Available from your local bookshop, or direct from Chartwell-Bratt, Old Orchard, Bickley Road, Bromley, Kent, BR1 2NE, UK. Tel 01-467 1956, Fax 01-467 1754